FROM AGRICULTURE
TO AGRICOLOGY

FROM AGRICULTURE TO AGRICOLOGY

TOWARDS A GLOCAL CIRCULAR ECONOMY

Professor
Dani Wadada Nabudere

MAPUNGUBWE
INSTITUTE FOR STRATEGIC REFLECTION (MISTRA)

MAPUNGUBWE
INSTITUTE FOR STRATEGIC REFLECTION (MISTRA)

Mapungubwe Institute for Strategic Reflection (MISTRA)
First floor, Cypress Place North
Woodmead Business Park
142 Western Service Road
Woodmead 2191
Johannesburg

© Prof Dani Wadada Nabudere (text)
© MISTRA (publication)

ISBN 978-1-920655-19-8

Published by Real African Publishers
on behalf of Mapungubwe Institute for Strategic Reflection
(MISTRA)

First floor, The Mills
66 Carr Street
Newtown, Johannesburg 2001

REAL AFRICAN PUBLISHERS

Editor: Angela McClelland

Indexer: Jackie Kalley

Printed and bound in South Africa

Contents

Foreword   11

Introduction   14

Agriculture in History   18

Modern Industrial Agriculture   26

Impact of Capitalist Industry and Finance on Agriculture   32

Agricultural Knowledge, Science and Technology   39

[A] The Green Revolution in Agriculture   53

[B] From Green Revolution to Gene Revolution   61

     i. Biology and Eugenics – a reductionist manipulative 'science'   62

     ii. Control of food markets and populations   70

     iii. Green and Gene Revolution in Africa?   79

     iv. From fossil fuels to biofuels – agrofuels   90

[C] The Food Crisis and Land Grabs in Africa   95

[D] Agriculture and Climatic Change   100

[E] The Destruction of the Small Farmers   106

[F] From the Old Industry to the 'New' Bio-industrial Economy   113

     i. A 'new science' for a 'new system'   113

ii. The 'new' industrial revolution   121

[G] Glocal Political Implications of the Agricultural Crisis   132

Towards a New Epistemology of Agricology   138

[A] Indigenous Knowledge Systems and Farmer Innovation   140

    i. General   140

    ii. How IKS is generated and innovated   143

    iii. Some case studies   146

[B] Mastering the Urban and Rural Divide   151

[C] The Resilience of Ecosystems and Small Farmers   161

[D] Re-conceptualising the Circular Ecosystems   167

[E] Restoring Traditional Governance and Justice   179

[F] From Agriculture to Agricology   182

    i. The 'science' of agroecology   182

    ii. Organic agriculture vs. Green Revolution   187

    iii. A transformative energy system   193

    iv. The seed as money   196

[G] Afrikology and Agricology   198

In Conclusion   206

References   213

Index   216

In Memory of Aduso –

The ancestral spiritual domain and custodian of the

Indigenous Knowledge of the Iteso people in

Eastern Uganda.

– Dani Wadada Nabudere

15 DECEMBER 1932 – 8 NOVEMBER 2011

In Memorium

Prof Dani Wadada Nabudere:
A champion of African self-reliance

By Jeffrey Sehume

It is a discomfiting irony that a continent as rich in mineral and human resources as ours manages to retain so few of either. On 8 November, Professor Dani Wadada Nabudere, one of Africa's finest scholars, who chose to remain on the continent throughout most of his life, passed away at the age of 78.

In his native Uganda, Professor Nabudere was admired as an academic activist, a former cabinet minister and a democratic educator with socialist leanings. To the global academic community weaned on his African-centred scholarship, he was a fountainhead of wide learning. The full extent of his significance to academia and society is still to be appraised. However, what is undeniable is the fact that Africa has lost an intellectual elder whose collected works span virtually the entire range of postcolonial independence social discourse.

Professor Nabudere was a singular individual, who qualified as a barrister in the United Kingdom in 1963, and who mastered the breadth of classical training in the social sciences. For him, education in Africa required synthesis with public needs in order to have lasting importance. Spurred on by this conviction, Nabudere dedicated his life to applying and spreading the notion of 'community sites of knowledge', which simply means using indigenous tools of knowledge to revitalise the lives of Africa's people. He staunchly believed that the liberation of Africans depended largely on self-reliance.

Nabudere was the Minister of Justice in 1979 and Minister of Culture, Community Development and Rehabilitation in 1979-1980 in the UNLF Interim Government of Uganda. He was also a former Associate Professor at the University of Dar es Salaam in Tanzania and at the Islamic University of Uganda, and a visiting professor at the University of Zimbabwe.

For Nabudere, dependence on imported knowledge and material

instruments could only lead to internalisation of colonial complexes, which dictated that ideas and knowledge that emanate from the West are superior to those that originate from the continent. Nabudere held firm to a confidence in African solutions to historical and structural problems.

To him, it seemed counterproductive to maintain the language of inclusion and exclusion inherited from colonialism. His commitment to life-long learning underlined his faith in the value of indigenous knowledge. He understood that African indigenous knowledge carries in its DNA the roots of 'complex ecosystems' that require the inputs of a diversity of expertise and experiences.

In this monograph, he wrote: 'This knowledge is ancient and has been continuously utilised for many millennia throughout the continent. For example, a team of American scientists doing research in Kenya in 1978 uncovered an astronomical observatory on the edge of Lake Turkana in Kenya, which they dated to 300 BC. According to them, it resembled the ruins of the Stonehenge in Scotland with huge pillars of basalt like the stumps of petrified trees lying at angles on the ground. Similar findings have been made among the Dogon of Mali in West Africa. All this goes to prove that indigenous knowledge in Africa incorporates scientific knowledge right from ancient times, which knowledge has been transmitted from generation to generation through the "Living Word" and "languages of the people".'

As principal of the Marcus Garvey Pan-Afrikan Institute (MPAI), he remained committed to empowering marginalised rural communities with tools for their own liberation as expressed in the term 'agricology'. Speaking at the launch of the Mapungubwe Institute for Strategic Reflection (MISTRA) in Johannesburg in March 2011, he defined the purpose of agricology as being to 'create local capacities for survival through local knowledge and languages'.

Nabudere's interest in alternative methods of addressing the world's problems was informed by a Pan-African and transdisciplinary perspective that sees the famine crises in Somalia as conducive to the rise of terrorism, piracy and threats to world peace. His research reflected an appreciation for the long-term impact that the 1884 Berlin Conference had in its creation of arbitrary national borders and the resulting conditions for ethnic posturing and conflicts between nation-states on the continent. The Berlin Conference has remained an albatross that continues to hang around Africa's neck, he argued. Of course, Nabudere was not averse to pointing out internal factors that continue to inhibit the realisation of African dreams, such as ineffectual leadership, or 'political godfatherism', as coined by another great African intellectual, Chinua Achebe.

Nabudere's sudden death brings home the meaning of the African proverb: 'when an elder dies a library burns'. The loss is all the more profound when the elder is someone who valued close collaboration between countries irrespective of ideology, geography, and technological advantage. His efforts at the institute he headed were focused on forging links and networks among organisations across the continent. At the time of his death, he was actively collaborating with the University of South Africa, African Institute of South Africa, and MISTRA. Nabudere recognised that even though the past could not be altered, it was nonetheless essential that energies were spent on reclaiming the future by producing knowledge that is relevant for society, and for the continued participation in civic causes designed to assist the wretched of the earth.

Nabudere is on par with irreplaceable sages like the Burkinabe historian, Joseph Ki-Zerbo, historian Bernard Magubane, and the nonconformist anthropologist, Archie Mafeje.

Nabudere died just ten days before he was to embark on a trip to South Africa. As we mourn the loss of a great African, we can take comfort from the fact that his collected works will continue to inspire new generations of African scholars who will take pride in affixing their own work to the experiences of the lives of ordinary people. We may now not be in a position to access new insights from the 'burned library' of his prolific mind, but Nabudere's intellectual legacy transcends material monuments and is assured to continue shining as a resource in the world's body of knowledge.

PROF DANI WADADA NABUDERE:
15 DECEMBER 1932 – 8 NOVEMBER 2011.

Jeffrey Sehume is a Senior Researcher at the Mapungubwe Institute for Strategic Reflection.

FOREWORD

This monograph is dedicated to a Community Site of Knowledge by the name PKWI Community Initiative; whose struggle over a period of 20 years has been aimed at upholding and emancipating their heritage enshrined in their indigenous knowledge systems, which were attacked by the colonial regime, but which they have been striving to recover and restore in their renewed attempt to create a new local 'green' circular economy. This knowledge was degenerated and criminalised so that members of their community who dared to propagate it, and/or dare and try to practise it for their self-sustenance, were accused of practising superstitious and backward ideas and were subjected to prosecution under colonial penal laws. However, the members of this community did not give up the struggle to preserve what was rightly theirs. They organised to rejuvenate the knowledge, which the colonialists thought they had destroyed. But because indigenous knowledge is permanent and self-renewing, the community has renewed itself by utilising it to develop their new economies utilising both the inherited knowledge and new knowledge derived from a holistic science they have been able to mobilise with partners.

The indigenous knowledge systems prevailing in this community include traditional scientific knowledge such as astronomy, which older women of the clans are experts in utilising up to this time. This knowledge is ancient and has been continuously utilised for many millennia throughout the continent. For example, a team of American scientists doing research in Kenya in 1978 uncovered an astronomical observatory on the edge of Lake Turkana in Kenya, which they dated 300 BC. According to them, it resembled the ruins of the Stonehenge in Scotland with huge pillars of basalt like the stumps of petrified trees lying at angles on the ground. Similar findings have been made among the Dogon of Mali in West Africa. All this goes to prove that indigenous knowledge in Africa incorporates scientific knowledge right from ancient times, which knowledge has been transmitted from generation to generation through the 'Living Word' and languages of the people.

Using this knowledge, the Iteso women in the PKWI Community Site of Knowledge are able to observe stars and trace their movements in a beer calabash, which enables them to predict the weather patterns to be expected in the coming months and thereby determine which varieties of millet or sorghum should be planted in the coming season with very good results. However, the colonialists and the colonisers could not accept such knowledge, which was barred from practise. In enforcing this, the British colonisers targeted a strong woman advocate of indigenous knowledge called Aduso, who was the custodian of the Iteso cultural spirits. Using their inherited knowledge, she fed all the lactating mothers on very nutritious indigenous root plants called *Ikorom* and leaves of a tree called *Edusa* (Moringa). The colonisers regarded these products as a source of superstitious, backward practises and criminal ideas, which opposed 'modernity' and colonial agricultural policies.

Instead, the colonial administration promoted the introduction of the Cassava crop in Teso in 1946, which they regarded as a more 'civilised crop'. Aduso's agricultural knowledge and practises and ideas were demonised, criminalised and abolished by the colonialists in that year. Their law enforcers devised a song that abused, demonised and belittled Aduso. In the song, Aduso was abused as being stupid, backward and barbaric. To spread their message faster, this song was taught in all Teacher Training Colleges and Primary Schools in Teso, and the neighbouring areas, so that the teacher trainees and students would propagate it further by teaching it to other children as innocent agents in killing their own ancestral heritage and knowledge. The author of this monograph can attest to the fact that he was one of those non-suspecting children who were taught this abusive song against African knowledge systems. The song went as follows:

Ebanga Aduso; [Stupid, Aduso],
Chorus; *enyami Ikorom, Aduso* x2 [She eats the Lilies-Ikorom, Aduso],
Aduselena; abotirana; ebangana- Aduso [She is fat and smiling, Aduso],
Chorus; *enyami Ikorom* x2 [She eats the lilies].

The elders, on realising the motive of the colonialists and their law enforcers, decided to preserve the Edusa plant but had its name changed to *Elekumare* (which means 'bring out the cow'), which the law enforcers could not know as they were not able to either identify or differentiate between the plant and the new expression, thus preserving the plant and the indigenous knowledge up-to-now. Since then, the two plants have been validated as very nutritious by modern science, thus demonstrating once more that Indigenous Knowledge Systems (IKS) are permanent and self-validating and incorporate holistic scientific knowledge, which does not disturb nature.

The persistence of this community in conserving and preserving their inherited heritage has enabled them to survive and continue to apply their knowledge and resist impositions such as the current Monsanto campaign to impose genetically-modified seeds and plants which are continually forced down the throats of those communities on the false claims that Genetically-Modified Organisms (GMOs) will help the community to combat hunger and global warming. Instead, the community has been able to engage universities and researchers to challenge their experimented 'scientific' knowledge systems. At one point, they succeed in forcing the universities of Makerere in Uganda and Ohio State University in the USA to validate indigenous knowledge systems concerning striga – the millet and sorghum parasite plant – and its control. This validation disproved the 'scientific' theories of the universities, which were being propagated. Since then, the community has continued to engage universities and government research centres to advance their indigenous ideas and practises, as we shall see in the body of the text of this monograph.

It is also important to record that the monograph was first written as a paper for presentation at an agricultural conference at the Walter Sisulu University in Mthatha, Eastern Cape, South Africa in 2011. However, it proved difficult to present the paper because of the formatting of the conference, which demanded more dialogue than the academic approach adopted in the paper. I therefore decided to enlarge the paper into a monograph, which we now present for general readership.

This monograph is therefore also presented in recognition of Professor Luswazi's efforts to establish a faculty of Agriculture and the Centre for Rural Development at the Walter Sisulu University in Mthatha, Eastern Cape, South Africa. Her work has also shown how the university can work closely with communities to promote a new agriculture for the future, which we have called agricology in this monograph.

Mbale, 14 June 2011   Prof Dani Wadada Nabudere

INTRODUCTION

The crisis of the on-going global economy, which has contributed to climatic change and the devastation of natural resources, including land all over the world, has begun to transform our understanding of the role industrial capitalism has contributed to this process. Agriculture has always been a fundamental economy on which humanity has survived throughout the centuries. However, the modern system of industrial economic management under capitalism has undermined the vitality of the soils on which agricultural production and human, plant and animal depended. Humanity is now called upon to revive the soils and restore the basis of human survival through a new system of production, nurturing, and caring if we are to survive the consequences of the present exploitative system. This system extracts from the earth all the gifts of nature without paying the replacement costs of its exploits. This irresponsible system can no longer be maintained.

Humanity cannot survive without nature nor can nature exist without humanity. There is a co-relationship between them and this co-relationship needs to be restored through a *circular global agricological system and economy*. The objective of this monograph is to add voice to this growing debate about the challenges facing humanity and what can be done to overcome the engulfing crisis. The neglect and exploitation of the ecosystem by modern society's search for private profit that does not compensate has the consequence of breaking down the unity and the interdependence of a coherent natural system. It should be realised that the ecosystem, which is built on the geological formations and the living forms that inhabit it, is a *single community of interconnected beings*.

This community includes the sun, the moon, the stars, plants, animals and humans who exist together as a compact unity. Each mode of existence of each of these entities has its unique rights and functions within the totality; and each is able to contribute to the existence of the other in a mutually supporting manner, and thereby to the maintaining of an integral functioning of the planet. As Thomas Berry has remarked:

> 'We are so related that the well-being of each member of the community depends on the well-being of the other members of the community. This law

of the integral functioning of the Earth constitutes what might be considered an ontological covenant bonding the universe into a single manifestation of the wonders of existence. This unity finds one of its highest expressions in the ever-renewing cycle of the biosystems of the planet [Milani, 2000: ix].'

Berry regards the last two decades of the twentieth century as constituting the period when humanity moved from a land-based, organic, ever-renewing economy to an extractive industrial, non-renewable economy [Ibid.]. He refers to an *industrial complex* of petroleum, steel, lumbering, pharmaceutical corporations and public utilities as being responsible for this transformation. This industrial establishment assumed the mandate to take over the resources of the entire globe to exploit for private profit without the consent of all humanity. Their exploitations have caused immense global devastation, creating extensive toxicity as well as leaving vast amounts of non-disposable waste materials. By exploiting the planet in this manner, a large proportion of the natural 'outer' world has been destroyed. The *outer world* has been so extensively affected that it has also affected the *inner world* of human experience because both these worlds 'are two dimensions of a single reality'. Berry adds: 'If an economy devastates the natural world then neither the psychic nor the physical structure of the human component of the Earth community can function in any satisfying manner [Ibid: x].'

As the industrial world reduces the Earth to lunar conditions, the humans also lose 'the sublimity of our inner lives'. We cannot, therefore, maintain a viable, sustainable economy by merely 'mitigating' or 'adjusting' to the consequences of the industrial devastation while at the same time maintaining the exploitative systems. A completely new attitude and a new mystique will be needed, which does not look at the Earth as mere 'resources' to be exploited. To move away from the old ways of living we have to adopt new human technologies, which are in accord with, and not subversive of, nature's technologies. These human technologies must integrate with the technologies of nature: 'drawing off what is appropriate for ourselves, yet remembering always that we must not deny other modes of being the fulfilment of their needs' [Ibid.].

Consequently, a new ethics is required, but the ethical value systems will not come from the heavens. The ethical system will have to come from the highest spiritual foundations of our ancient heritage, which has always contained ethical systems guiding the relationship between human beings and other forms of existence right from the Cradle of Humanity in the Nile Valley. These systems have been subjected to change but their basic and core values still remain. Many scientists today are pursuing what is called a 'convergence' movement, which they regard as being an answer to the present crisis caused by industrialism. Convergence in mainstream scientific systems aims at identifying a definite value, a definite structure, a common view or opinion, or something toward a fixed or equilibrium state.

Recently however, the concept has been widened to include the questioning of the contemporary separation of knowledge into compartments called 'academic disciplines'. This questioning has arisen out of our common understanding of the changes that have taken place in science, which now recognises that science largely relies for its subject matter on a common knowledge of the ordinary people about things. Concepts of life and death, plant and animal, health and sickness, youth and age, mind and body, machine and technical process, etc., are commonly known without the intervention of science. It is from these sources of inspiration, combined with our memories of a more moral heritage from the past, that we can draw our ethical relations with nature. As Professor Cheikh Anta Diop has rightly put it:

> 'A new ethics that largely takes into account objective knowledge ... and the interests of the human species is in the process of being built; it is only difficult to internationalise it because of the conflicts of national interests. Ecology, defending the environment, tends to become the foundation of a new ethic of the species, based on knowledge: the time is not too far off when the pollution of nature will become a sacrilege: a criminal act even, and mainly for the atheist, because of the one fact that the future of humanity is at stake, what knowledge or 'science of the epoch' decrees as harmful to the whole group thus becomes progressively a moral prohibition [Diop, 1980: 375].'

Since the global capitalist crisis of 2007-2008, there have been attempts by transnational corporate bodies to transfer the effects of the financial aspects of the crisis to poor countries, especially, as we shall see below, the grabbing of vast swathes of land from small farmers in Africa for biofuels, which will intensify the 'monetisation' of biomass production and its conversion to industrial uses. Attempts are also being made to grab vast areas of land by financiers holding vast amounts of toxic wealth, such as derivatives, in order to buy land cheaply and use it as a safe haven for their toxic assets, especially if they are able to invest and produce large profits of say a 400 per cent return on their investment in a short time. This will intensify the use of genetically-modified, engineered methods of production and increase the use of new kinds of destructive fertilisers as well as terminator seeds in countries where the GMO seeds had not yet spread.

This will add to the pressures on the fertility of the land. It will also fuel deforestation, especially in the geographical areas of the world that are the world's reservoirs of biomass, leading to greater climatic warming, which will further weaken the soils and their capacity to produce food. This drive will require a resistance to ensure that the remaining parts of the world that have not been destroyed by industrial agriculture are saved from further destruction.

AGRICULTURE IN HISTORY

According to Sir James Frazer in his book, *The Golden Bough* [1922, 1966], agriculture had its origins in ancient Egypt and was connected with the myth of Osiris, the most popular of all Egyptian deities. According to this myth, Osiris was regarded as a personification of the great yearly vicissitudes of nature, especially corn. He was also connected with the invention of the calendar of 365 days through which nature produced and reproduced itself through human action and natural processes and *in a circular and recurring manner of birth and death*. The myth goes that Osiris had a sister called Isis whom he married and who was his companion. Reigning as the first king of Egypt, Osiris reclaimed Egypt from savagery and gave the Egyptians laws, which taught them to worship the gods.

In this process, it is said, Isis discovered wheat and barley growing wild and that Osiris then introduced the cultivation of these grains among his people who abandoned cannibalism and took to the corn diet. Osiris is also said to have been the first to gather fruit from trees and to train vines to poles and to tread the grapes. Eager to communicate these discoveries to all mankind, Osiris is reputed to have left the whole government of Egypt to his wife, Isis, so that he could travel all over the world to diffuse the blessings of civilisation (including agriculture) to the rest of mankind wherever he went. The myth goes that wherever he went and found the soil unsuitable and the climate harsh to the cultivation of the vine, he would teach the local inhabitants to console themselves for the want of wine by brewing beer from barley, instead [Ibid: 436-37].

The myth also goes that Osiris was known as the Corn god so that one of his personifications was corn, which was said to die and come to life again every year. This fact demonstrates that even at this early stage, the circular nature of agricultural production was part of the mythology of humanity and the basis for human action in agriculture. The myth was, in fact, manifested into reality through rituals that celebrated in festivals the birth and resurrection of Osiris as personified in the corn from which the Christian belief of resurrection was drawn. The festival was celebrated in the month of *Khoiak* and at a later period in the month of *Athyr*. According to Frazer:

'That festival appears to have been essentially a festival of sowing, which properly fell at a time when the husbandman actually committed the seed to the earth. On that occasion an effigy of the Corn god, moulded of earth and corn, was buried with funeral rites in the ground in order that, dying there, he might come to life again with the new crops. The ceremony was, in fact, a charm to ensure the growth of the corn by sympathetic magic, and we may conjecture that as such it was practised in a simple form by every Egyptian farmer on his fields long before it was adopted and transfigured by the priests in the stately ritual of the temple [Ibid: 454].'

Therefore, before modern science and through a belief in sympathetic magic, the ancient Africans believed in the circular character of natural reproduction and this 'green-ness' was also inserted in the myths of the time. The myth had to do with the understanding of fertility, which was associated with green-ness. Osiris was also personified as a Tree spirit and in celebrating that fact, every year a pine tree was cut down and its centre was hollowed out and, with the wood thus excavated, an image of Osiris was made out of it, which was buried like a corpse in the hollow of the tree. In this respect, Osiris was also personified as a god of fertility. In this role, Osiris was conceived as a god of creative energy in general, which in its ritual form was believed to be a charm to ensure the growth of crops. Frazer conjectures that Isis, the wife and sister of Osiris, must have been also called the Corn goddess. This was why Isis was referred to by the Greeks as the 'many-named', and the 'thousand-named' [Ibid: 460].

Being the one who discovered wheat and barley, she was, according to Augustine, associated with the myth of Ceres. She was also referred to as the 'create-ress of green things' and the 'green goddess whose green colour is like unto the green-ness of the earth', a 'Lady of Bread', and a 'Lady of Abundance'. The Greeks also conceived Egyptian Isis to be a corn goddess and identified her with their own goddess called Demeter. In their epigram, they described Isis as: 'she who has given birth to the fruits of the earth', and 'the mother of the ears of the earth'. It was through these co-options of the myths of Osiris and Isis by the Greeks that Ceres was regarded as the goddess of grain who introduced grain-based agriculture into Europe.

Agriculture was therefore one of the arts of civilisation that the *Genesis of Osiris* brought to the world as we have seen. The English word *cereal* is derived from *Ceres*, the name of the goddess of grain, according to the Greeks. In European conventional mythology, Venus and Ceres are both regarded as 'Roman' goddesses. However, as Frazer points out, they were conquered goddesses because the Genesis of the Egyptian Osiris and Isis preceded the Roman Empire by several centuries. At the time when grain culture was first introduced in Europe, Rome was just a rustic frontier settlement. Therefore, Venus and Ceres are both older than the Roman Empire that claimed them as Roman goddesses. It also follows that both Venus and Ceres are younger than Osiris and Isis, from whom they are created. Whether or not either Venus or Ceres were Latin goddesses before they were Roman goddesses is not clear, but it is also immaterial; they may both have been Mother goddesses, who were gift wives of conquered peoples, and this conquest must have been at the hands of the Egyptians.

This is because, according to Professor Diop, the predominance of the Black civilisation around the Mediterranean region is attested to by the 'unexpected existence of the pre-Hellenic Black virgins and goddesses, such as the 'Black Demeter of Phigalia in Arcadia, the Black Aphrodite of Arcadia and Corinth, the Black virgin of Saint Victor of Marseilles, and the Black virgin of Chartres, who was once honoured as Our-Lady-under-the-Earth [Diop, 1980: 21].' Diop concludes, therefore, that 'the cult of Black virgins, which the Church finally sanctified in modern times, derived directly from the cult of Isis, which preceded Christianity in the Northern Mediterranean [Ibid: 22].' Diop adds that we lack scientific proof to connect them to the Aurignacian Venus, 'but their existence confirms the southern (i.e. African) origin of civilisation' [Ibid.].

Thus, from the very ancient times, people belonging to different civilisations, through cross-cultural sharing and learning, had a similar view of nature and its products such as agriculture, fisheries, animal husbandry and forestry. They also regarded these products of nature, the animal world and labour as being the product of a *circular process* in their very connection with nature and the reproduction of food crops and animals for human consumption. The production systems were connected to moral and ethical belief systems that guided the thinking and actions of the people. In reviewing our relationship with Mother Earth in the current crises that confront us, we need to revisit some of these central messages from the origins of humanity and their connection with nature, in order to restore the balance between humans and other members of the natural community if we are to survive as a collectivity.

Western scholars have been divided on the origins of agriculture. Due to the racist 'Aryan Model' of the origins of civilisation [Bernal, 1987], many of these scholars have argued that agriculture in antiquity originated in Mesopotamia, which they believe was one of the first places where plant domestication occurred. They argue that following the domestication of plants and animals, numerous settlements developed in this region and add that archaeological evidence suggests that the first city, Jericho, was built in the area. They point out that the most important crop plants for all of these civilisations were (and still are in that region) wheat, barley, dates, figs, olives, and grapes, and merely guess that it is possible that each of the crops were domesticated from wild plants native to the region.

However, they recognise that the most advanced civilisation of the region, however, arose along the banks of the Nile River where they admit the first exotic ornamental gardens were established. They point out that the Egyptians deliberately planned the grounds and placed plants for both ornamental and utilitarian purposes. They date the first records to about 2200 BC, which is rather too late for discussing the origins of agriculture. They nevertheless point out that among the elements typical of some of the gardens were: pools for fish, trees bearing figs, pomegranates and dates, grapevine-covered trellises, and beds of flowers. Other plants included roses, jasmine and myrtle. The gardens were primarily those of pharaohs and government officials; and those that served as religious or sacred sites.

However, other scholars such as John Claudius Loudon in his *Encyclopaedia of Agriculture* [1825] have stated the historical record more accurately. In the book he correctly traces the origin of agriculture from Africa (i.e. Egypt) and summarises the historical record as follows:

'Traditional history traces man back to the time of the deluge. After the catastrophe of which the greater part of the Earth's surface bears evidence, man seems to have recovered himself (in our hemisphere at least) in the central parts of Asia, and to have first attained eminence in arts and government, and on the alluvial plains of the Nile (River). Egypt colonised Greece, Carthage, and some other places on the Mediterranean Sea; and thus Greeks received their arts from the Egyptians, afterwards the Romans from the Greeks, and finally the rest of Europe from the Romans. Such is the route by which agriculture is traced to our part of the world; how it may have reached eastern countries of India and China, is less certain, though from the great antiquity of their inhabitants and governments, it appears highly probable that arts and civilisation were either coeval there, or, if not, that they travelled to the East fully more rapidly than they did to the West [Ibid: 3].'

Thus, James Frazer's conjectures about the origin of agriculture dating back to the days of Egyptian Osiris and Isis is better grounded, and it is from here that we can correctly assess the temporal and spiritual basis of agriculture, forestry, fisheries, and animal husbandry as natural circular economies. From here we can see how agriculture began and then spread to the rest of humankind, and how in each region of the world different kinds of crops and seeds were discovered and managed. Indeed, recent archaeological discoveries of African agriculture reported in the British science magazine, *Science*, in 1979 revealed that the earliest leap from hunting and gathering activities to the scientific cultivation of crops occurred in Africa at least 7000 years before it appeared on other continents. The magazine reported discoveries by one, Fred Wendorf, of agricultural sites near the Nile going back to more than 10 000 years before the dynasties of Egypt were founded. In these areas, Africans were cultivating and harvesting barley and einkorn wheat as well as other crops. These were carbon-dated at Kubbaniya, a site just a few kilometres north of Aswan, and the dating gave a reading of 17 850 to 18 000 years ago. Other sites further south, like Tushka in Nubia, indicated the same dating.

These discoveries revealed that not only were the Africans the first to engage in crop science, but they were also the first to domesticate cattle. A University of Massachusetts anthropologist, Charles Nelson, reported in *The NewYork Times* in 1980 that his research team had unearthed evidence in the Lukenya Hill district in the Kenya Highlands about 40 kilometres from Nairobi, the capital, that Africans had been domesticating cattle some 15 000 years ago. According to the story:

> 'Nelson said that the finding led them to conclude that pre-Iron Age Africans in that area had a relatively sophisticated society and could have spread their mores, living modes and philosophy, eventually reaching the fertile crescent of the Euphrates River Valley, which many had once thought was the cradle of civilisation. African technologies, he suggested, were exported to the Middle East through trade and the cultural diffusion of information and ideas [von Sertima, 1999: 322].'

But the diffusion of African agricultural technology was not limited to the Nile and the Euphrates regions. Some African-cultivated crops such as a strain of African cotton, as well as a species of jackbean and yams, were reported in the American agricultural complex before Columbus, while African indigenous rice, finger millet, sorghum, pearl millet and cultivated cotton had entered western Asia very early. The question arises: How did these technologies and crops spread to other major agricultural complexes? According to Ivan von Sertima, as the Sahara became drier from 5020 TO 2500 BC, Africans were forced to migrate to other parts of the continent, 'taking their crop science with them' as they moved from one part of the continent to the other [Ibid: 323].

That is how we are able to infer the wide and early distribution of language networks and shared agricultural terms, linked to cultural and techno-complexes, and the study of plant geography. This is also what has made it possible for scholars such as Professor Clyde Winters, based on his studies of agricultural complexes in West Africa, to demonstrate in several academic papers the movement of Africans in Western Asia and the African origins of the Dravidians in India. It is the Dravidians, belonging to the C-group in Nubia, who have been credited with the diffusion of several African domesticates to Asia. Professor Winters has established many connections between Dravidians, who are now Afro-Asiatic, and black Africa, particularly through linguistic evidence and shared cultural traits. He also demonstrated that the crops mentioned earlier, which were indigenous to Africa, appeared later in the Asian context in the areas where the Dravidians had settled [Ibid: 324]. Ivan von Sertima has also, in the case of Latin America, demonstrated in great detail in his book, *They Came Before Columbus*, that a number of crops of African origin had found their way to the Incas. From this evidence Sertima came to the conclusion that 'Africans indeed had the most intimate knowledge of plants, a fact used to great advantage not only in crop cultivation but also in the development of medicine [Ibid.]'.

This historical record shows that agriculture has indeed a longer history than previously imagined by Western academic research based on their pre-determined diffusionist theories which favoured Asia and the Middle East as centres of ancient civilisation over Africa. Archaeological research carried out on the African soil has revealed otherwise. It also demonstrated why the modern capitalist agricultural theories based on reductionist scientific techniques are indeed short-sighted. They have transformed the entire agricultural economy in the developed western world, and the world over, to individualistic capitalist interests, which have informed the motive for the colonisation and exploitation of the land and the crops that they selected for agricultural expansion for world markets without regard to the prior circular principles on which agriculture was based.

Small farmers in India after the devastations of the Green Revolution always asked the scientists the question: 'from where did the notion that unless plants had something constantly "done to them" they could not grow and not produce, come from, and creep into modern consciousness?' The answer seemed to have been provided by Albert Howard, the founder of modern organic farming in America, who discovered that the entry of modern chemicals into agriculture on a sustained basis occurred when companies manufacturing explosives and weapons of World War II found themselves without a market or a niche for their products once the world war was over. This is how the use of organo-phosphorus insecticides based on the poisonous properties of phosgene gas discovered during WWII came into the picture for commercial purposes. Howard also drew attention to the fact that the same corporation that produced Agent Orange gas chemicals that were used to defoliate the tropical forest in Vietnam during the Indo-Chinese war fought by the Americans in the 1960s, was the very corporation that manufactured herbicides today that kill plants, herbs, weeds and seeds not desired by agribusiness. Alvares concludes from this that:

> 'Modern agricultural theory conveniently reduced the plant to a combination of nitrogen, phosphorus and potash (NPK), (and thus) laying the ground for a massive diversion of the deadly production of these war industries and their chemicals to agriculture [Ibid: 9].'

He adds that agriculture has never been the same since this intervention. Even with the destruction showing clearly throughout the world, agribusiness today, with the active support of governments, is continuing to expand their control over agriculture across the world, taking it out of farmers' hands and continuing to mess it up. They are doing this by promoting a Green Revolution for Africa, pushing GMO seed to small farmers in Africa 'free of charge' as an inducement, and grabbing lands for biofuels and 'food security' as an agribusiness and thereby placing the entire humanity at a risky dependence on food produced by agribusiness.

These practises have resulted in the ecological devastation that has been taking place gradually over the last 200 years since the emergence of industrial capitalism. As Berry has remarked above, it is especially over the last two decades of the twentieth century that the process has been intensified by what he called the 'industrial complex' of transnational corporations and public utilities, during which period we have moved from a land-based organic, ever-renewing economy to an extractive industrial, non-renewable economy. This negative impact of the extractive activities of these corporations has moved from the application of organic fertilisers to the more intensive use of synthetic fertilisers, insecticides and herbicides. This has happened through phases as the crisis in agriculture also intensified.

Modern Industrial Agriculture

The above historical understanding about the origins of agriculture through mythology and history was changed with the rise of the modern world economy under capitalism, which began to structure new social relations on land and its connection with nature. According to the German philosopher and economist, Karl Marx, the genesis of the capitalist farmer had its origin in a slow process, which evolved through many centuries to a fully-fledged system. According to him, in Europe the serfs held land under different tenure systems and were therefore emancipated under very different economic conditions. In England, the first form of the farmer was as a *bailiff*, who was himself a serf. This was similar to the condition of a *villicus* in ancient Rome. These relations were altered as the economic and political conditions changed.

In Europe, in the second half of the fourteenth century, the serf was replaced by a feudal peasant farmer, whom the feudal landlord provided with seed, cattle and farm implements to produce for him. His condition was not very different from that of the earlier serf, except that the feudal landlord exploited his labour power more intensely than was the case with the earlier serf peasant under slave conditions. But soon the feudal peasant farmer became a share-cropper. He entered into a contract with the feudal landlord and divided with the landlord the product in proportions that were determined by the contract. This relation was soon replaced in the case of England by a capitalist farmer proper who raised his capital by employing wage labourers on a permanent basis after the Enclosure Movement.

The peasant serfs were dispossessed by a process of this movement and turned into paupers – who provided capitalism with a 'reserve army' of the unemployed who were to become wage labourers. The capitalist farmer was able to extract a surplus product from the labourer from the reserve army of the unemployed. This surplus product was in kind, or in money-form, and from this surplus the capitalist paid the landlord ground rent. As capitalism developed, this surplus product was increasingly paid in money form and became 'surplus value' [Marx, 1976: 905-8]. But during the fifteenth century, there were still pockets of independent peasant farmers and farm labourers who worked for themselves, as well as for wages, thus enriching themselves by their own labour.

The agricultural revolution, which began in the last third of the fifteenth century, changed all this as we have already seen above. The usurpation of common lands enabled the emerging capitalist farmers to augment their stock of cattle, almost without cost. They were able to fertilise the land from the supply of cattle manure for the enrichment of the soil, thereby producing higher yields. During the sixteenth century, the progressive fall in the value of precious metals (i.e. gold and silver coin), and therefore of money, resulted in the lowering of wages of labour, which added to the profits of the capitalist farmers. This also led to the rise in the prices of corn, wool, meat and all other agricultural products, which swelled the money capital of the capitalist farmers without much action on their part.

This was also at a time when the ground rent the capitalist farmers were contracted to pay to the new capitalist landlords did not rise but on the contrary diminished since it had been contracted on the basis of the old values of money. This resulted in the capitalist farmer becoming richer at the expense of both the labourers and the landlords, leading to the consolidation of capitalist farming in England at the end of the sixteenth century. Without going into what happened in other countries in Europe, Asia, Latin America and the United States of America, it is important to note that this transformation on land revolutionised all relations of production in the rest of the economy, and changed the character of the agricultural economy, increasingly, all over the world.

This revolutionary transformation led to the industrialisation of agriculture, while agriculture also revolutionised the manufacturing industry by supplying it with new products in the form of raw materials, food and animal products based on capitalist economic and social relations. This made agriculture become an essential part of manufacturing. The raw materials that had hitherto been produced as part of 'indigenous agriculture' now became part of the elements of 'constant capital', alongside machinery, chemicals and other products, which manufacturing produced. The product that was created out of this combination of 'free' labour, raw materials and machinery, became a total 'commodity capital' that was sold to realise *profit* for the industrial capitalist who became the 'captain of industry'. This is the same process that also created the home market and ultimately led to the emergence of the international market on the basis of

European imperialist expansion all over the world.

Marx had also incisively examined the impact of industrial production and distribution on agricultural production. In a chapter dealing with the time of production in Volume II of *Capital*, Marx observed that in the new system, the production time necessary for the advanced capital consisted of two periods – the working period and the production period. In the case of agriculture, the production period was subdivided into two sub-periods: one consisting of the *labour process*, while the other consisted of *the time when capital is in the form of existence of the unfinished product*. In this period, the unfinished product 'is abandoned to the sway of natural process, without being at that time in the labour process': unlike in industrial production proper. Therefore, in these two sub-periods, the working time and the production time do not coincide. In this case, the production period is longer than the working period. In the extra period, the product is not finished and not ready to be converted from *productive-capital* into *commodity-capital*, ready for marketing [Marx, 1956: Vol. II: 243].

In case of the extra production time, inasmuch as it is not fixed by natural laws such as the growing of an oak tree, the period of *turnover* can be shortened by the *artificial reduction* of production time. This can be done by the use of chemicals or other artificial additives such as fertilisers. Marx added that the difference between working time and production time becomes especially apparent in agriculture, where the period of production depends on the alteration of good and bad seasons, as well as on the quality of the soil. The more unfavourable the climate is, the more congested is the working period, and hence the shorter the time in which capital and labour are expended. In case of the soil, poor soil has to be enhanced by artificial chemical fertilisation or natural fertilisers such as cow dung. All these require additional capital, which makes agriculture dependent even more on industry for its inputs, making agriculture a subsidiary element to industry and the farmer an employee of the industrial capitalist.

Therefore, the key element in this revolutionary change was the transformation of land into *landed property*. Land had now *a price* so that the general prices of production of agricultural commodities were now determined by the cost of production on the *worst* piece of land plus an average rate of profit. This was a radical departure from price determination in the manufacturing industry, where the price was based on a social average of cost of production that prevailed in the entire sector of the economy. This departure in price determination was justified on the basis that the 'naturally based' differentials in productivity on land could not easily be eliminated by technological change in the same way it could in industry. It was also argued that an expansion in agricultural production entailed drawing more inferior lands into cultivation as well as intensifying production on superior soils when it was more profitable to do so [Harvey, 2006: 342].

This mode of price determination also affected the way ground rent was fixed. The important thing to note was that since land (or the earth) was not a product of labour, it did not have a money 'value'. Therefore, what was sold or 'rented' as *landed property* was a claim by the landlord to receive annual revenue from the capitalist 'renter' or buyer, which for the purposes of industrial production assumed the form of 'fictitious capital'. The land became a form of fictitious capital and was treated as a 'pure financial asset', which was bought and sold according to the rent or 'price' it yielded. Therefore, ground rent was in effect a claim on future profits from the use of land, or more directly, a claim on future labour, employed by the capitalist farmer.

But the emergent capitalists on their part strived as much as possible to make profit without paying rent, and hence the tendency on their part was to move from 'old' land where *absolute rent* was payable to the landlord, to 'new' lands where the capitalist could reap profits without rent. Here the capitalist state dispossessed the peasants and it, itself, became the landowner from which it dispensed titles to the capitalist farmers and corporations free of rent. This explains why, at a certain stage, the capitalist imperialist states resorted to the colonisation of peasant lands in the southern part of the world, not only as part of the 'primitive accumulation' of capital, but also for the production of food and raw materials for industry at home.

According to Marx, the essence of a 'free colony' consisted of the fact that the bulk of the soil had become 'public property', and therefore every settler on it could turn part of it into private property. This meant that individual capitalists were able to turn the 'free lands' into a means of production for themselves without preventing future settlers from having the same benefit [Marx, 1976: 934-8]. Various mechanisms were used to turn the colonised 'free' peasants and labourers into wage labourers in order to fully exploit colonial land as a capitalist landed property. But this was done before the free peasant or 'indentured labourer', too, could earn enough money to become a capitalist farmer on his own account. In the colonies that were conquered later, such as those in Africa, the peasants resisted dispossession, except in southern Africa, where land and cattle were taken by force. Even here, the laws of capitalist production on land were resisted, and a resort was made to turn resisting peasants into bonded labourers. As Professor Mafeje observed:

'The white landlords, whose estates were measured in 1000s or at least 100s of acres, behaved more as feudal lords than as a capitalists, and treated their African labour as bonded labour, which was often paid in kind. This took the form of rations and squatting rights in exchange for labour, which in principle was available for 24 hours a day and extended to the members of the worker's family [Mafeje, 1988: 99].'

Thus, under these conditions in Africa, laws were put in place by the imperialist states, such as the South African Master and Servant Act of 1845, and the use of 'traditional' systems, such as the *Tangatha* system in Nyasaland (Malawi), under which the bonded workers had no right to withdraw their labour under the contract or system. This was the only way the capitalist could maintain landed property as a 'fictitious capital', which could be traded just like any other commodity based on labour exploitation in conditions where the land itself had no value. In this way, it was the capitalist that created modern agriculture and brought it under the control of industry in different environments. It is with the same process that *wage labour replaced the real earth as the ground on which society stood*. It also became the reason why landed property led back to wage labour, from which it extended all the way from the cities to the countryside and thereby distributed over the entire surface of society [Marx, 1973: 276-8].

In conditions where there was a preponderance of cheap labour, especially in the colonies, this fact tended to slow down the introduction of technology in agriculture. This was true of the situation in Africa, where in the case of South Africa, the availability of cheap, captive labour, according to Mafeje, 'considerably slowed down technological progress in agriculture in Southern Africa' [Ibid: 99]. Otherwise, the application of science in the production process in agriculture for the first time in history became possible leading to the full development of the productive forces in agriculture under capitalism in different parts of the world. This presupposed the total restructuring of the agricultural mode of production and therefore it also presupposed certain conditions which rested on the development of industry, trade, and of science, as 'the forces of production' [Ibid: 277]. Even the development of manufacturers presupposed the dissolution of the old economic relations on land. But it was also true that agricultural production was less productive than industrial production, and hence the need for competition on one side, and, on the other, the development of science in the form of chemistry and mechanics as products of the manufacturing industry to move agriculture as a partner to industry. It was not surprising, therefore, that in the twentieth century there was an enhanced productivity in agriculture through the use of synthetic fertilisers and pesticides, selective breeding, mechanisation, water contamination, and farm subsidies in order to increase its 'productivity'. It is also not surprising that it is this same process that has led to the degradation of the earth as a product of nature.

Impact of Capitalist Industry and Finance on Agriculture

The crisis of capitalism in industry, which underlies the industrial agricultural crisis in general, is at bottom an industrial crisis of loss of profitability in manufacturing. Capitalism has long exhausted all possibility of profitable production and reproduction except in a few countries with a low composition of capital. This aspect is expressed by Marx as a long-standing 'tendency of the rate of profit to fall' inherent in capitalist production. Marx had proved theoretically that it was only human *labour power* that was capable of producing wealth in the form of 'surplus value'. This was possible because the capitalist had managed to dispossess the peasants of their landed property and turned them into a labour force, which the capitalists could exploit through the system of wage labour. Marx called the wages 'variable capital' and the capital in the form of machinery which he deployed to extract labour power 'constant capital'. He argued that machinery or technology ('constant capital') did not produce surplus value. Marx predicted that this combination of constant capital and variable capital would eventually lead to the 'tendency for the rate of profit to fall over time', as the capitalist increased machinery and reduced labour due to higher wages, which the workers were able to obtain through trade unions to increase all the time [Marx, Ibid.].

This was due to the fact that in order to remain profitable and competitive, all capitalists had to invest more and more in technology in industrial production in order to increase their 'productivity', since an increase in wages payable to workers was bound to lead to a drop in the rate of profit in general. It was necessary to reverse the process of investment by investing more in constant capital and less in variable capital, but this action of the capitalist merely intensified the contradiction because it had the paradoxical effect of enhancing the tendency of the rate of profit to fall even further. The same law applied to the agricultural sector because machinery in the form of tractors, chemicals, fertilisers and herbicides were all products of industry, which were applied in agriculture.

Marx also observed that this tendency of falling profitability was the same tendency that led to the *concentration* of capital in fewer and fewer corporations in the form of trusts and cartels. This was because as companies became fewer and larger, it was possible for them to fix monopoly prices in order to increase profits in money form. Therefore, Marx predicted that capitalism's inherent tendency to concentrate the ownership of the means of production over time would weaken it over time as labour also became better organised to demand better working conditions and wages in the work place. Therefore, there would be a constant struggle between the two classes for 'privatisation' and 'socialisation' of production.

In order to overcome this tendency, the capitalists were forced to *centralise* capital by merging industrial capital and banking capital into what Lenin called 'finance capital'. This enabled them to privatise all the profits from industry and the banks by also 'owning' the deposits of savers. By creating banks, the finance capitalists acquired the power to concentrate all savings into a pool of capital, which they could lend freely and obtain interest from it. They paid only a small portion of this interest to the savers and pocketed the rest in the form of 'expenses', 'costs' and 'fees', which formed part of the 'dividends' which they paid to the capitalist 'shareholders'. In effect, they became real owners of the peoples' savings. This centralised wealth was in effect grabbed by an 'aristocracy' of finance capital, which emerged out of this cartel of monopolists in industry and the banks. But this power could not have been achieved without the support of the State in their operations through legislation, etc. This is why, since the end of the Second World War, there has been a general decline of material production in industry in the capitalist world and an increase in the financial sector and banks from which a greater part of their profits were obtained. This has resulted in a resort to 'financialising' as a means of 'making money out of money without production'. Instead of production of material goods, the capitalists have increasingly resorted to the production of paper, which they call 'money' or 'credit'.

This is a sign that capitalism can no longer activate human labour into producing material goods, and hence resorts to speculation in 'financial instruments', which has also affected industrial agriculture, since agriculture, as we noted above, had become an aspect of capitalist industrial development. Profits in agriculture are also increasingly being obtained from speculation in food and agricultural materials.

The speculation in financial activities and instruments, instead of production of material goods and services, has been exacerbated by the power of investment banks to 'leverage' and create new credit instruments such as 'collateralised debt obligations' – CDOs and 'derivative contracts' to represent 'wealth', without corresponding material goods. Therefore wealth has become ownership of paper wealth with a minimum of production of luxury goods. This is what produced 'financialising' as an 'economy' whose 'products' were 'sold', 'exchanged', and 'banked' just like material commodities, leading to an inverted pyramid created by various 'Ponzi schemes' that manufactured money. This has also affected agriculture, especially food production and exchange. As we shall see below, certain food products were financialised to become 'hedges' and 'back-ups' against losses in the 'value' of the worthless 'paper wealth' in financial instruments held by investment banks. That is why the economic 'meltdown', which occurred from September 2008 to date, has affected certain food products, which have been used in financial 'hedging' [Nabudere, 2010].

The crisis of the agricultural economy was especially felt in the increase in the prices of key food products such as the grains, soya beans and fruits. The fall in the productivity of industry on which agriculture depended meant the collapse of the agricultural raw material economy as well. This is because agricultural production in these products could not be maintained without the demand from the industrial sector. It is this decline that affected agricultural production as well, but certain agricultural and food products were focused on for purposes of backing up the financial 'services'. Therefore, those food products were targeted for financial speculation as they were hoarded in order to provide a hedge against fluctuating money values created by the financial instruments.

The speculation in food products took the form of 'future trading' and 'trading in options' as a form of 'commerce'. That is why financial instruments, especially 'future options' and instruments called 'derivatives' on food trading continued to grow in volume while production of these food products declined in volume. In short, in order to maintain 'stable' prices in the targeted food products, production had to be 'managed' and 'priced' at certain levels in order to ensure their 'profitability' to the financial instruments. As a result, those selected food commodities became major 'hedges' as a back-up to the financial 'future contracts' and derivatives, which were expressed in US dollars. To ensure their proper marketing, the trade in these products was handled by the centre of the global commodity trade based in Chicago called the Chicago Board of Trade (CBOT). It is here that global trade in commodities is valued, managed and exchanged in coordination with other commodity markets in other centres of the capitalist world, such as London. It is also here that all commodities, including food commodities, are financialised into dollar financial instruments created by the financial sector.

Particular food commodities were targeted for this role; these included: wheat, oats, corn, rice, and soybeans. Other products included beans, oil, coffee, cocoa, sugar, cotton and orange juice: all 'soft' commodities, many of which were traded on the Coffee, Sugar and Cocoa Exchange (CSCE). The number of financial future contracts for wheat, which were made as commitments to buy or sell a given volume of wheat at a certain future date at a predetermined price, for instance, had quadrupled over the five years (2003-2008). What made matters worse, especially from 2007 onwards, was the fact that speculation had become especially active around the record low inventories in the existing agricultural commodities, especially the above food commodities. In the past 25 years, most countries had gradually abandoned the policy of stocking grains and other agricultural commodities, which acted as food security stocks due to the unpredictability of production and the high cost of storage. With the food market in turmoil, there were no food cushions to absorb the impact of any sudden disruption in the ability of the commodity markets to import grains into the major commodity markets. This pushed food prices upwards.

During the first three months of 2008 – the year the global economic crisis intensified, international nominal prices of all major food commodities reached their highest levels for 50 years. The United Nations Food and Agriculture Organisation (FAO) reported that food price indices had risen, on the average, by eight per cent in 2006 compared with the previous year. In 2007, the food index rose by 24 per cent compared with 2006, and in the first three months of 2008, it rose by 53 per cent compared with 2007. This sudden surge in prices was led by increases in vegetable oils, which on the average increased by 97 per cent, followed by grains with an increase of 87 per cent, dairy products with 58 per cent and rice with 46 per cent.

The sudden rise in prices of these commodities was caused by the deregulation of financial markets and the systematic exploitation of US regulatory loopholes by the investment banks. This had led to the upsurge of speculative investments in food commodity markets, much of it by 'institutional investors', such as the managers of pension funds, as well as the investment banks. These banks and funds had ploughed tens of billions of dollars into the targeted agricultural commodity 'assets' as a way of diversifying their asset base and improving 'returns' for their investors. But it is clear that the economic malaise was much deeper than these appearances and it was this malaise in the 'real economy' that led to the relaxing of regulatory controls to enable these fund managers to better manage the food market themselves without bureaucratic controls and supervision. But as has recently been revealed, banks and other financial and commodity corporations had spent billions of dollars bribing the legislators not to legislate against their speculative activities in these products.

For instance, a 231-page report entitled: *Sold Out: How Wall Street and Washington Betrayed America* [2009], revealed that the financial sector spent over $5 billion in buying political influence in Washington over the decade by employing as many as 3,000 lobbyists to do the buying of politicians to 'waive and deregulate' the regulations that had been put in place to contain speculative activities in food products. The report documents a dozen regulative moves that were waived, that ultimately led to the financial meltdown, and which cultivated a 'culture of recklessness' and greed. It was this private pressure that contributed to the repeal of the Glass-Steagall Act of 1933, which President Roosevelt brought into legislation to deal with the speculative effects of the 1929-30 financial crises.

The Glass-Steagall Act had prohibited commercial banks from offering investment banking and insurance services to the financial sector, thereby ensuring the separation of the commercial banking and investment banking. This separation was now removed by the Financial Services Modernisation Act of 1999, which made it easier for investment banks to act without restraints in promoting financial instruments that had no material backing of any kind except the hedges that were provided by these selected food products. This enabled the banks to 'invest' monies from individual depositors' cheque and savings accounts into 'creative' financial instruments such as mortgage-backed securities and 'credit default swaps'. These were the investment gambles that rocked the financial markets in 2008.

Because agricultural markets are small, relative to stock markets of industry, the amounts of cash that were pouring into these markets gave the speculative banks a substantial clout in their speculative activities, especially in the selected food products. The big institutional investors such as mutual funds and investment banks controlled enough wheat in futures instruments to supply the needs of American consumers for a few years. When the crisis struck, the blame for the 'demand shock' that had arisen was placed on the 'recent (speculative) entrants' to the commodities markets as the primary factor behind the sudden soaring of food prices. It was noted by independent observers that if no immediate action was taken, food and energy prices were bound to rise still further, leading to the catastrophic economic effects on millions of already stressed US consumers, and the possible starvation of millions of the world's poor.

The key breakthroughs for the speculative investors came in the energy futures market, where general precedents for all commodity trades (including food) were established. In 1989, the US Commodity Futures Trading Commission (CFTC) issued a policy paper in which it declared that it would not regulate 'swap deals', meaning commodity purchases involving financial intermediaries, which typically were investment banks and the other commodity dealers. This was followed in 1990 by a declaration that the Commission would consider oil trading on the Brent Market as 'forward contracts', or a contract in which the buyer takes delivery of the commodity immediately, rather than 'futures contracts', or contracts in which the buyer rarely takes delivery, but instead uses the contract for purely speculative purposes.

Analysis revealed the extent to which agricultural production had been undermined, which led to the emergence of financial and commodity markets resorting to speculative 'investment'. This development also indicated the limits to which capitalist industrial agriculture had reached. The speculative capitalist economy in agriculture had not only undermined the material production (in agriculture and industry) in monetary and financial terms, it had also undermined the sustainability of agriculture as an industry. The crisis had hit hardest at the level where people's lives were critically affected, namely in the food production sector.

In the poorer countries, such as those in Africa where capitalist agriculture was not developed to the fullest extent, indigenous crops, which also served as medicinal plants, were undermined by their replacement with imported exotic crops such as wheat, rice and other crops that were consumed by the colonial and later post-colonial elites. The forested areas that were home to many of these crops and plants were also cut, thereby destroying the basis of their existence. This created conditions for hunger in these countries even when the countries were fertile. They also lost the capacity to produce different kinds of food crops, fruits and vegetables. In the case of the pastoral communities, the colonial policy prevented pastoralists from moving around, and instead made laws that compelled them to settle and engage in sedentary production. This enforced settlement resulted in the over-grazing of lands, especially when modern drugs were introduced to fight animal diseases. This had the consequence of increasing the numbers of cattle, which in turn led to over-grazing, drought and desertification, resulting in the degradation of the environment and pastoral life.

AGRICULTURAL KNOWLEDGE, SCIENCE AND TECHNOLOGY

The underlying crisis that contributed to the food crisis, and agricultural production in general, as we have seen above, arose from the damage that had been caused to agricultural production by scientific and technological inputs, apart from the market factors. This means that the agricultural knowledge that had been developed and applied to agriculture as scientific knowledge was destructive of the natural regenerative processes rather than supportive of them. According to a report compiled by a team of scientists from international agencies entitled *International Assessment of Agricultural Knowledge, Science and Technology for Development*, and the accompanying global report entitled *Agriculture at a Crossroads* [2009], during the last 50 years the physical and functional availability of natural resources had shrunk faster than at any other time in history due to increased demand and/or degradation of nature at the global level. This was compounded by a range of factors including human population growth and its impacts, which has led to unprecedented loss of biodiversity, deforestation, loss of soil health, and pollution of water and air quality.

The report added that, while in many cases such negative impacts could be mitigated, given the multifunctional nature of agriculture, it was critical to consider links across ecosystems in which agricultural systems were embedded. These links had important implications for the resilience or vulnerability of the entire systems. The report observed that the linkages between natural resource use and the social and physical environment across space and time were an important issue for agricultural knowledge, science and technology, with significant implications for sustainable development and the mitigation of adverse impacts.

The underlying factors to the malfunctioning of industrial agriculture have especially to be traced to the role played by science and technology in the industrial agricultural economy from its inception. This is because without the application of science and technology agricultural production, as we know it today, could not have emerged. Here consideration has to be given not just to the 'scientific paradigms', which Thomas Kuhn demonstrated to be rather short-lived, but to the epistemological foundations on which these paradigms themselves are based. Behind these epistemologies, we have also to trace the cosmologies and 'mythoforms' that underlie their worldviews. These are the foundations that inform the scientific epistemologies and paradigms as well as the technologies as understood and applied today in all sectors of life, but which undermined the earlier non-European epistemologies.

Steve Fuller [1998] has observed that before the modern Western epistemologies about agriculture developed their scientific methods on the basis of the native disempowerment, the non-European communities such as the Incas of Latin America had advanced their knowledge of agriculture far ahead of them. According to him, the disempowerment was made through an attempt to 'rewrite the histories of the non-European peoples' when these peoples were not around to criticise the histories constructed on their behalf by the Europeans and could not, therefore, critique them. He gives the example of the Incan settlement at Machu Picchu in Peru, which archaeologists now believe to have been an agricultural station devoted to the study of the effects of different soils on crop yields. According to Fuller, the level of soil discrimination evidenced at Machu Picchu by the combined efforts of botanists and chemists in Europe until the mid-nineteenth century could not match them. Indeed, he adds, European scientists had not been driven to acquire agricultural knowledge with the Incan eye for detail 'until potatoes imported from the Americas helped to eliminate many nutritional disorders in Europe'. This lengthened the lifespan, which, coupled with a constant fertility rate, led to an increase in the rate of population growth in Europe. But as we noted earlier, the knowledge about these crops and soils had been diffused from Africa to the Americas.

As a result, it became important for Europeans not simply to grow more potatoes and kindred crops, but also to determine how crop yields could be increased per unit of arable land. Experimental investigations along these lines had clearly been undertaken at Machu Picchu five centuries earlier. Fuller observes:

'However, because the Incas demonstrated their agricultural knowledge 'merely' in experimental practise, and not as deductions from any recognised principles of natural science, Europeans have tended to downgrade the Incan contributions as 'pre-theoretical', 'trial and error' ('inductive', if one is being extremely polite), or 'technology' (as opposed to 'science'). Each of these expressions is meant to suggest that had the Incas 'really' understood what they were up to, they would have eventually arrived at the theories of agronomy that we now regard as providing the best explanation for the level of agricultural success they achieved [Ibid: 88-89].'

Fuller notes that as military history has come to be fitted into the history of science, 'Europeans have reluctantly come to admit that their own *conquistador*-like urges may have contributed to the Incas' failure to extend their agricultural knowledge beyond a certain point. A still bolder, more under-determinationist hypothesis is that the Incas had a radically different sense of the social role of knowledge in their world, one that the visiting Europeans could not readily grasp, given their lack of schooling even in their own scientific traditions'. Fuller observes further:

'In view of their different cosmology and world-view, which did not recognise a unified cosmos under one creative deity, the European scientists were at a distinct disadvantage, given their Western tradition of treating plants as degenerate animals and biochemical changes as 'emergent' on underlying physical processes. This is simply to say that agriculture significantly deviated from the paradigm of the Western scientific inquiry, mathematical physics [Ibid: 89-90].'

Thus we come to understand where the Western scientific paradigms began to depart from those of the non-Western sciences, which predated them. The Cartesian epistemology is based on the belief of such a creative deity who existed outside the universe in a *logico-mathematical* world, which is a creation of the European Enlightenment. It is an evolutionary and unilinear epistemology according to which mind and body are separate entities in which the mind, and not the heart, is the creator of knowledge. It is an epistemology that is based on mimicking and imitation of nature for the purposes of its control and exploitation and not in accordance with the laws of transformation of matter. The maxim for this 'new' knowing individual is 'I think, therefore I know!' The 'I' of the new era was unlike the 'I' of the previous era, which Levy-Bruhl had characterised as a 'participation mystique' where the state of mind in which the 'I', the individual entity, did not have sharp boundaries. The 'mystical participative' entity is merged in his environment, so that the consciousness and world in which humans exist are deeply intertwined. In this cosmology and worldview, the consciousness is not an exclusive attribute of the thinking 'subject', but an attribute that belongs to everything, and is everywhere, and according to it: 'The whole world is animated and has an anima, a soul [Sabbadini, 2010: 1-2].'

The new 'I' entity was an individual unto him/herself – a producer of knowledge. The Cartesian scientific 'method', which the thinking subject used in 'knowing', was based on a philosophical conception called *naturalism*, which emerged with the Enlightenment. This epistemology resulted from what Charles Taylor [1985] called 'a shift in cosmology' among the protagonists of the Enlightenment movement. The shift dropped the old natural law doctrine according to which all things had a permanent and unalterable structure that existed in accord with universal and immutable laws. The shift enabled the new epistemology to 'objectify' everything into independent situations that could only be known by investigation and 'proof'. This was necessary in order to overcome the weaknesses of the old world-views and cosmologies and their ways of knowing that were now regarded as either 'erroneous', superstitious', or 'backward' [Taylor, 1985: 2-4, 242-44]. These cosmologies were ecologically-friendly and not expansionist.

This shift in cosmologies necessitated the creation of 'sciences' such as the *natural sciences*, which became 'the most successful application' of research. The new reductionist epistemology based on 'naturalism', according to Taylor, was part of a metaphysical 'motivation' which was 'many-faceted', but one way of defining it was the status accorded to the natural sciences as the models for the 'sciences of man' (i.e. social and human sciences). In this metaphysical contrivance, 'man' was seen as being part of nature, but in a manner in which he, too, along with nature, could only be understood 'according to the canons which emerged in the seventeenth century revolution of natural sciences' [Ibid: 2].

'One of the most important of these (canons) is that we must avoid anthropocentric properties ('subjective properties'), and give an account of things in absolute terms. 'Anthropocentric' properties are those which things have only within the experience of agents of a certain kind – the classical example in the seventeenth-century discussion are the 'secondary' qualities; while 'absolute' properties are supposedly free of any such relativity. This requirement can be more or less stringently interpreted and can be applied at different levels, which accounts, I believe, for the variety of reductionist views, but it underlies all of them [Ibid: 2-3].'

Therefore, this 'reductionist' epistemology of regarding 'man' as part of nature was itself restricted in that man as nature could only be comprehended and guided by the 'new laws' which were to be devised by some 'men' who had the correct capacity to know and produce knowledge. The 'scientific method' would explain human beings just like other objects in nature. This meant that all 'anthropocentric' phenomena derived from the five senses, such as colour or felt temperature, and those based on human feelings, were to be disregarded as unreliable 'secondary properties'. Emphasis was to be placed only on 'absolute properties', which would 'account for what happens invoking only properties that the things concerned possessed absolutely ... that is, properties which they possess even if they are not experienced' [Ibid: 242-3]. Taylor observes from this that:

'Hence we get a demand which is widely recognised as a requirement of materialism in modern times: that we explain human behaviour in terms of goals whose consummations can be characterised in physical terms. This is what, for example, for many Marxists, establishes the claim that their theory is a materialist one: that it identifies as predominant the aim of getting the means to life (which presumably could ultimately be defined in physical terms) [Ibid.].'

But this mode of explaining the matter can be obscurantist, for it is not Marxism that is responsible for the materialist approach in its creation. Most significantly, the materialist orientation arises within the class and religious circles that believed in the *immateriality* of god but were able to reduce capitalism into a materialist goal, which Christianity, especially Protestantism and Calvinism, played a significant role in establishing. Christians, and not Marxists, were the leading materialists, who turned Christianity into 'Secularism' in order to see capitalism as a 'spiritual' system for satisfying material needs, but on a 'progressive' and expanding basis with no end, ostensibly ending up in heaven or hell.

This is what Steve Fuller [2010] has called 'Protescience'. Fuller argues that our continuing faith in science in the face of the crises that are occurring all around us is due to the 'secular residue of a religiously-inspired belief in divine providence'. He believes that this originated in the Abrahamic religions:

> 'It goes back to the biblical Abraham, who was driven in old age and without offspring, by a voice he took to be from God, to leave the land of his birth and settle in unknown land, where he proceeded to prosper and even bear children. However, that same voice then told him to sacrifice his male heir, Isaac, which he followed once again until it (the voice) told him at the last minute that a ram would suffice. From that point onwards, the path to human redemption began to be charted. The three religions that have been most responsible for the rise of science – Judaism, Christianity and Islam – all claim this story as their founding moment [Fuller, 2010: 1, 6-7].'

Fuller reads two important implications from this narrative. The first was the belief that the future will be better than the past, 'indeed so much better that it led him to stake just everything on it'. According to Fuller: 'This sort of brinkmanship, whereby individuals willingly sacrifice themselves and each other for a "promised land" no matter what the cost, really comes into its own with the deployment of science as a vehicle for modernising the world' and all that flows from that belief [Ibid: 7]. Perhaps the best illustration of that brinkmanship, apart from the Dr Strangelove story which Fuller describes, is the Al Qaeda belief that by sacrificing oneself to blow up and kill innocent human beings, the person sacrificing oneself will go to heaven straight away and be rewarded for his 'holy' act.

The second implication of the Abrahamic narrative is that which concerns the deity in whom Abraham began to believe. According to Fuller, who has done some extensive research on these issues, this is a moral problem because Abraham 'ends up finding a proxy sacrifice' for his son 'in a way that enables God to test Abraham's faith, but at a much lower cost to all concerned'. Although Abraham sees this act of his God a merciful act, his God may not have viewed the matter in the same way:

> 'It is more plausible to suppose that God, having realised that Abraham was already inclined to do his bidding, judged that Isaac's sacrifice was unnecessary and that perhaps sparing Isaac would have the added benefit of stiffening Abraham's resolve in the future [Ibid: 14].'

Fuller argues that the implication of this is that by ascribing to God such a 'cunning sense of reason', as Hegel had called it, Abraham had succeeded in turning the divine intellect of God into 'an extended version of our own in a manner in which humans became "masters of the Earth"'. This is the 'spirit' which science came to derive from this understanding so that what scientists later promised was a mode of being which could be called 'divine', 'in which we are all endowed with a much greater storehouse of information; capacity to act on the world; powers of prediction and, perhaps most importantly, universe of concern; specifically one that extends not only to the normally foreseeable future but to all time': this is what gave us the power to talk of 'progress', or the capacity to 'improve'.

In short, we 'sublimated' God as an end of organised enquiry and 'secularised' Him to enable us to transcend our normal modes of experience: 'enquiry into which is never to be discouraged, despite the regular occurrence of both cognitive error and political failure'. Fuller asks why we so believe in such transcendence. His answer is: 'Because scientists and their well-wishers presume that the effort will be rewarded in the long term; *we are fallible but not corrigible.*' This faith in the power of scientific enquiry would be arbitrary, according to Fuller, without the Salvationist sensibility imported from theology into science. This would also explain the scientists' 'concerns for detecting and mastering divine agency', a concern of the so-called realist approach to science. It may also explain our acceptance of the biblical saying that we are 'made' 'in the image and likeliness of God', known since St. Augustine as the *imago dei* doctrine [Ibid: 15].

Following these transcendental premises, the new epistemologies constructed under the all-powerful naturalist cosmology demanded that all paradigms in the natural sciences be mathematically determined, and that it was only through this path that 'true' knowledge could be acquired, tested and validated to be the basis of predicting events based on that knowledge. This new approach required that 'all statements with claims to truth must be public, communicable, testable, and capable of verification or falsification by methods open to and accepted by any rational investigator'. This was because mathematical methods were regarded as being the most reliable means of recording, predicting, and therefore controlling, nature, whose real character remained inscrutable [Berlin, 1979: 1-2, 162-3].

The men of the Enlightenment, therefore, rejected all other forms of authority; and in particular such foundations of authority based on religious faith and sacred texts, divine revelation and/or 'dogmatic pronouncements of its authorised interpreters', except in the way they themselves interpreted them and relied on them. They also rejected tradition, prescriptions, immemorial wisdom, private intuition, 'and all other forms of non-rational or transcendent sources of putative knowledge', although on many occasions they also relied on them. Berlin adds:

> 'This principle was held to apply to both to the human and non-human world: abstract disciplines, such as logic or mathematics, to the applied sciences which established the laws of the behaviour of inanimate bodies, plants, animals and human beings, and to the normative disciplines which revealed the true nature of ultimate human goals, and the correct rules of conduct, public and private, social and political, moral and aesthetic [Ibid: 163].'

The new epistemology therefore banished the earlier beliefs, which linked human action to any moral or metaphysical systems of knowledge. The scientific approach dropped any notion of a circular renewable energy force and any circular natural economy, and instead adopted a nineteenth-century unilinear evolutionary theory of reality, according to which all human societies had evolved along a common path from simple hunting and gathering communities to literate civilisations. In this new epistemology, all societies were supposed to pass through the same basic sequence of development or evolution or 'stages of development', although the speed of transition might vary from stage to stage. This was a single line evolution, which did not allow for diversity, including biodiversity and circular developments. It allowed only one version approved by science, and a model, which is behind the contemporary 'one-size-fits-all' dominant ideology of capitalist exploitation of nature and human labour based on wage and 'market' slavery.

It was with this epistemology that modern theories of technological development and application were crafted and promoted. Since reality consisted of a unilinear path, all technological development had to be developed through this unilinear evolutionary process and adopted everywhere. This is what led to the idea of the *Transfer of Technology Model*, according to which technological knowledge in the most developed part of the world had to be 'transferred' to the less developed, or the 'undeveloped world', for them to 'catch-up' with the more developed world. This was, in fact, a reversal of the evolutionary theory itself, since evolution was supposed to be from the bottom upwards. But the reversal worked quite well for the new era, which promoted capitalist development to occur all over the world in the same way.

This model, however, contradicted the human historical experience based on diversity. For that reason it has led to a one-sided, one-centre mode of development which is lop-sided and destructive of nature and human labour. The model has failed to meet a broader range of development goals that address the multiple functions and roles that agriculture has historically played in the diverse agro-ecosystems. According to the *Agriculture at a Crossroads* report [IAASTD, 2008], under this model, science and technology have been mobilised under the control of experts who define the problems and then design solutions to them in a unilinear evolutionary fashion. Even then, the scientists selectively tap some of these old systems of knowledge mainly for local adaptation purposes, where this is considered desirable and profitable to the 'market'.

But these changes must be understood in their political contexts of European imperialism and its expansionism into non-European societies. This political order is what fashioned the way science was crafted to serve the dominant economic interests of the ruling classes of the imperialist countries. This worldview can be traced to the origins of Greek classical thought, which later came to inform another direction that Christianity came to have impact and influence on science. According to this Greek worldview, as expounded by Aristotle, the universe was organised in a divine hierarchy over which a supreme being, God, presided, and in which the rest of the creatures followed in a descending pecking order. In the pecking order human beings, especially men, ruled over all the creatures below them, which comprised women, children, koala bears, snails, phytoplankton, and rocks. This Aristotelian scholasticism carried all the elements of expansionism that came to characterise the modern western civilisation with its roots in Christianity and Judaism.

In this way of looking at the world, Aristotle argued that the conquest of 'natural slaves' was right, and therefore war against them was justified. These thoughts later became central doctrines of the Judeo-Christian-Islamic religious traditions, in which the natural environment existed mainly for the purposes of meeting human needs of the superior conquerors. Christianity became inherently expansionist in practise and philosophy. By destroying paganism, which was closely linked to nature, Christianity made it possible to exploit nature in a mood of indifference to the feelings of the natural objects. Animism that characterised paganism was substituted by the cult of saints, which was functionally different from animism under paganism. But the saint was not in natural objects and, although he may have special shrines, his citizenship is in heaven. But he is entirely a man who cannot be approached in human terms. In addition, there are angels and demons who were inherited from Judaism and perhaps also from Zoroastrianism and are also mobile as the saints themselves. This had the fateful consequence that the spirits which existed in natural objects, and which protected nature from man, evaporated, and man's monopoly over spirits in this world was confirmed, and the old inhibitions on the exploitation of nature crumbled [White, L. 1967: 1202-1207].

Therefore, there emerged a two-tier hierarchisation of the universe. The first was that God, who was now externalised to be outside the universe, was supreme over all beings. The second was that humans, especially the men, were supreme over all other beings, beginning with women and children, who were considered inferior. The third was that 'natural slaves' could be enslaved permanently and, if necessary, killed in conquests carried out by the superior races. These hierarchies were advocated and practised by the Catholic Church in its political role, for instance in its conquests over Latin America. Another example is when citizens of Rome tried to organise protests over the slaughter of bulls for amusement and sport, Pope Pius IX refused to allow them on the grounds that animals had no souls, and therefore deserved no moral sympathies from man. Thus, despite some minority opinions, such as that of St Francis of Assisi, the Roman Catholic Church remains up to now ambivalent and indifferent about nature.

Therefore, all the imperialists needed to do (including the Church itself, which authorised the conquest over natural slaves in South America, Asia and Africa through the 'Bulls') was to argue that the races they were conquering were inferior in order for them to exploit and/or exterminate them. The philosophical and religious basis was laid for the expansionist imperialist ideology that ultimately led not only to the conquest and extermination of natives in the other continents, but also to a 'scientific' and mechanistic worldview which sanctioned the genetic expansionism. It is not, therefore, surprising that one of the earliest advocates of expansionism was Francis Bacon, whose exploits of science were now used to extend a 'scientific' system as a source of power and tools for the control of nature. As Calestous Juma has observed:

'Bacon stressed that for all their pompous claims, the Greeks had not performed any experiments which led to improvements in the human condition. For him, the main goal of science was to endow human life with new discoveries and powers. He advocated the search for objective knowledge which would enable humankind to have control over natural things [Juma, 1989: 42].'

This attempt to rationalise and expand outwards was intensified by Rene Descartes, who then proceeded to 'mathematise' all expressions and the behaviour of things through precise and neat mathematical laws. This was possible because Bacon had provided the basis through 'rationalism'. The Greek view of the world as a series of chaotic events and decay was deemed irrational and false. Juma points out:

'With the Cartesian method, the world could be reduced to separate entities which represented the whole; the behaviour of the sum of the parts was equal to the functioning of the whole. This mechanistic view of the world, which received new impetus from Galileo and Bacon, was now on its way to theoretical dominance [Ibid: 42-43].'

According to this approach, Descartes compared human beings and animals to machines such as clocks, springs and wheels. He defended vivisection and cruelty to animals on the ground that they did not deserve human compassion because their cries were mere creaks of wheels. Newton compounded this mechanistic view by presenting a methodological synthesis of the opposing empirical, inductive method of Descartes. For him, what mattered was the existence of some form of equilibrium in which different entities existed, and in which the entities remained in balance through mutual attraction. Gravity was his main scientific discovery, which he regarded as the ultimate force in the universe until Albert Einstein deconstructed it. This was because it was constant and the most fundamental of all laws. The universe was seen by him as a mega-clock, a God's handicraft, which was reversible since it was in equilibrium and was timeless. This view disregarded history, since all events occurred instantaneously and could be analysed through comparative statistics and since also all that happened was already determined in the original conditions. In this analytical scheme, all things changed through linear progression so that if there were changes, they were predictable, and the objective of research was to establish their cause-effect correspondences.

These mechanistic views about the universe had serious implications for the development of agriculture through the use of metaphors. For instance, the metaphors on biology, genetics and agricultural production led to the need for the classification of plants so as to understand more fully their potential contribution to human needs. This is because the external form of plants was presumed to be stable and discernible, and therefore 'an analytical method that could capture their key features could be used for classification' [Ibid: 44]. This ignored their internal characteristics. The belief was that plants were created by God in the form in which they existed, which made it easy to classify them according to their external distinctive morphological features, which was attempted by the Swedish botanist, Carl von Linné (Linnaeus).

The method Linnaeus utilised in classifying plants from their external morphological features was simplistically mechanistic, although it was appealing. According to the method one simply counted the pistils and stamens in a flower, for example, to determine its position in God's divine arrangement. With the use of the Cartesian mathematical calculations, the plants could be classified and neatly organised on shelves according to their pecking order in the arrangement. This method was adopted in most European countries which had 'successfully' synthesised and rationalised the existing methods of classification and analysis, and this was due, according to Juma, to the prevailing genetic expansionism. The Cartesian nature of the method was appealing to the scientific community in a period of abstraction and reasoning, which conformed to the growing need to base botanical studies on herbarium specimens, as it was becoming difficult to deal with large bodies of botanical information, 'some of which was on plants from all over the globe' [Ibid: 44].

In his 1749 book entitled *The Oeconomy of Nature*, Linnaeus reiterated the Judeo-Christian view that nature existed only for the purpose of meeting human needs, which was also clearly Newtonian. The species of nature in the economic system were cyclical in that they returned to the same point of departure with the precision of clockwork. He did not concern himself with variability within the species since these were internal to the plants. Thus with the application of science to the study of botany, humankind was able to utilise the natural endowments of each to satisfy their needs since God had set the limits to the geographical range of each specie and assigned its peculiar food. From this perspective, nature was seen by the British imperialists largely as a storehouse for raw materials for industrial and agricultural production, which influenced economists such as Adam Smith to look at nature as a 'resource' which was inexhaustible.

This was further supported by Newtonian equilibrium notions, which led to the view that all pollutants and waste released in the environment by industrial activities would gradually dissipate as the system returned to 'equilibrium'. The costs of ecological damage were regarded as externalities to the units of industrial and agricultural production. The challenge that came from Darwin's *Origin of the Species*, which questioned the whole idea of species being the handwork of God, but rather as a product of natural evolution and selection, was overwhelmed and incorporated into the scheme of Linnaeus of classification, and botanical research continued as business as usual, only reinforced by the discoveries of Gregor Mendel, whose studies pointed to the existence of 'units of heredity' that were later referred to as 'genes' [Ibid: 46].

With Mendel's work, the discipline of genetics was born as a realisation of the mechanistic and reductionist programme in biological sciences. Under the discipline, it was asserted that every gene corresponded to a specific trait in a linear way. The understanding of the functioning of the whole system was subordinated to the imperatives of the genes. This was used to promote 'genetic determinism'. Life forms were treated as machines controlled by linear chains of cause and effect. This deterministic interpretation of natural phenomena has been used to justify expansionist practises such as racism and sexism against Africans and other non-Western peoples. Genetic determinism was immediately applied to agricultural production with plant breeders that have been reflected in the emergence of monocultural agricultural production.

Monoculture is genetic expansionism with all the reductionist and mechanistic underpinnings, which led to 'new knowledge' about the functions of enzymes applied to industrial production, especially in fermentation. This became especially evident in the 1940s with the DNA reflected as the self-replicating molecules that carry genetic information and form on the basis of their chromosomes. Molecular biologists have since unravelled the basic language of life as coded by DNA so defined, enabling the scientists and industrialists to transfer specific genes from one organism to another. Since 1973, when these techniques began to be applied, scientists have not only been able to control and use life forms, they have also tried to modify and create new life forms out of these processes, including biotechnology, which has helped push this expansionism further, with untold benefits to the expansionists and risks to the generality of the human beings.

The mainstream expansionist reductionist model aims at achieving immediate impact, and in the case of agriculture it tries to do so by increasing productivity of the land by the use of long-term destructive technologies and external inputs. These strategies include functioning producer and service organisations, the social and biophysical suitability of technologies transferred in specific environments, and proper management of those technologies at plot, farm and landscape levels. In some cases, as we have seen, local adaptations are made based on traditional knowledge and innovation systems, often through participatory and experiential learning processes and multi-organisational systems in promoting innovation of diverse farm systems, but in a controlled manner.

The evaluation of the long-term impacts on agriculture of this scientific approach must therefore take into account such grave impacts as climate change, land degradation, reduced access to natural resources (including genetic resources), bioenergy demands, transgenics and trade requiring special efforts, and investments in agricultural knowledge, science and technology that relate to those local and global concerns. To do this, these different challenges that modern industrial agriculture has created have to be addressed in order to deal with their adverse impacts. In the sub-sections below we shall try to examine some of these scientific applications and assess their impacts before we can examine new ways of dealing with the crises that have been created by industrial agriculture under capitalist conditions. We shall restrict our concerns here to the following developments: [A] The Green Revolutions in Agriculture; [B] From the Green Revolution to the Gene Revolution; [C] The Food Crisis and Land-Grabs in Africa; [D] Agriculture and Climatic Change; [E] The Destruction of the Small Farmer; [F] From the Old Industry to the new Bio-Industry; and [G] The Glocal Political Implications of the Agricultural Crisis.

[A] The Green Revolution in Agriculture

The Green Revolution was introduced on the world scale in a number of countries with the purported objective of introducing industrial methods to achieve a greater productivity and fight 'world hunger'. The term, Green Revolution, was first used in 1968 by the former USAID director, William Gaud, who noted the spread of the new technologies and said: 'These and other developments in the field of agriculture contain the makings of a new revolution. It is not a violent Red Revolution like that of the Soviets, nor is it a White Revolution like that of the Shah of Iran. I call it the Green Revolution.'

But the origins of the Green Revolution are to be found in the United States' dominant families who were organised as economic corporations seeking to dominate world markets in food production, oil (energy) and pharmaceuticals. The chemical industry emerged out of the military operations of the Second World War (WWII) against Germany. The corporations that were involved in the manufacture of these destructive chemicals – DuPont, Dow Chemicals, Monsanto, Hercules Powder and others – were faced with a glut of nitrogen production capacity which had been built up during the war at the taxpayers' expense to make bombs and explosives against the enemy. Nitrogen was a prime component of TNT and was used in the manufacture of explosives. But it could also be used to form the basis of nitrate fertilisers for agricultural uses [Engdahl, F. W., 2007: 123-24]. Therefore, arising from this reality, the chemicals industry was able to develop new products to capture large markets for their nitrogen in the form of fertilisers, ammonia nitrate, and anhydrous ammonia for both domestic US agriculture and for export abroad.

According to Tim Jackson [1996], the creation of massive amounts of synthetic substances that neither natural systems nor our own immune systems can tolerate, has happened in a very short time. He points out that in 1900, even after 150 years after industrialisation, 'over half of the total materials in use (excluding those used for fuels and for food) were still provided by agricultural, wildlife and forestry products'. It was after the Second World War that an explosion in the creation of new materials based in petrochemical production occurred, leading to a rapid toxic pollution of agricultural production. By 2000, 40 per cent of toxic pollution from manufacturing was derived from the chemical industry, and this became strategic for the capitalist waste economy on which market 'demand' depended. Besides being a major polluter in its own right, the chemical industry provided a large portion of its end-products that were intrinsically damaging, which provided the means of destruction to other industries such as manufacturing, agriculture and forestry. Agriculture became the largest source of pollution due to chemical runoff; and was also essential for extending production and consumption loops, and in that way, in breaking down or distorting the natural circles of nitrate fertilising.

The nitrogen fertiliser industry, which began after the Second World War, was part of a powerful lobby of the Rockefeller Standard Oil circles, which by the end of WWII, included DuPont, Dow Chemicals and Hercules Powder, among others. The global market in the new agro-chemical industry after the war had also solved the problem of finding significant new markets for the American petrochemical industry as well as the grain cartel, a group of five companies, which included Cargill, Continental Grain, Bunge and ADM. Therefore, the largest grain traders in the world market were American, and their growth was a product of the development of special hybrid seeds, which were used through the spread of the Green Revolution in the 1960s and 1970s to expand their markets. According to Engdahl: 'Agriculture was in the process of going global and the Rockefeller Foundation was shaping the process of agribusiness globalisation [Engdahl: 124].'

With this dominance and monopoly of the agricultural chemicals and of the hybrid seeds, American agribusiness giants were preparing themselves to dominate the global market in agricultural trade. This was also part of a strategic aim of the US government, which had already in 1970 published a *National Security Memorandum-NSSM 200*, written by Henry Kissinger, the US Secretary of State, who propagated the control of food production and population control throughout the world. In his pronouncements Kissinger occasionally commented: 'If you control oil, you control nations', and 'If you control food, you control people.' This is what determined US foreign policy in creating food aid as part of its policy in the form of privately sponsored food agricultural inputs. This was aimed at controlling nationalist movements not to move towards socialism advocated by the Soviet Union in the mid-1960s. Therefore, the combination of US policy of offering food aid and the scientific techniques which were being developed in the name of the Green Revolution presented a golden opportunity in the global control of world food production as part of the strategic policy of US agribusiness and the US government against the USSR and 'communism' in general.

These developments laid the ground for the application of this knowledge and agricultural process to Latin American and Asian countries, as well as to Africa much later. This was the result of the judgement at that time that conventional industrial agricultural methods and techniques had not produced the necessary build-up in agricultural productivity to meet the galloping needs of rapidly growing populations. In the 1950s and 1960s there were fears of famine occurring in India and a number of other countries in Asia and Latin America. In fact, these 'fears' expressed by the promoters of the Green Revolution were generated as part of the strategy to introduce this form of agriculture on an experimental basis in developing countries, but with the design of promoting agribusiness and US global dominance of the food economy.

The beginnings of the Green Revolution are attributed to Norman Ernest Borlaug, but Borlaug was an employee of the Rockefeller Foundation and his work was that of advancing the Rockefeller global agenda in agriculture. In the 1940s, Borlaug was sent to Mexico by the Rockefeller Foundation where he began conducting research on seeds. He was able to develop new disease resistant high-yield varieties of wheat, which were not yet genetically-modified but which were to become so in the next phase. This was supposed to overcome what the Foundation regarded as the coming 'Population Monster' that would lead to environmental and social ills that he said too often led to conflicts between men and between nations [Borlaug, 1958]. But this was a disguise of what the Rockefeller Foundation had already begun in establishing the basis of its global business interests. Nelson Rockefeller was already an active and leading figure in US corporate investment in Latin America since the 1930s. In the 1940s their Foundation tried to set up the Mexican American Development Corporation through which they had made heavy investments in Mexican industry after the war. They set up the Chase Bank Latin American division as a way of gaining a foothold in Mexico through the disguise of helping to solve Mexico's food problem [Engdahl, 110-11].

By combining Borlaug's wheat varieties with new mechanised agricultural technologies, Mexico was able to produce more wheat than was needed by its own citizens, leading to its becoming an exporter of wheat by the 1960s. Prior to the use of these varieties, the country was importing almost half of its wheat supply [Borlaug, 1968]. In 1966 the Rockefeller

Foundation was joined by the considerable resources of the Ford Foundation and its intimate ties with the US government, US intelligence and foreign policy establishment to push the Green Revolution agenda outside the US. In that year, the Mexican government, alongside the Rockefeller Foundation, set up the International Maize and Wheat Improvement Centre (CIMMYT), which focused on wheat experiments originating from the breeding studies that were begun in Mexico in the 1940s by the Rockefeller Foundation.

This programme received a boost with the US government food policy known as P.L. 480. Under this policy it was decreed by the US government that no food aid was to be dispensed unless the recipient country had agreed to preconditions, which included agreeing to the Rockefeller agenda of agricultural development. This, in effect, meant that the US government was using food aid under P.L. 480 foreign policy to push the Green Revolution under the Rockefeller programmes as well as US investments and population control programmes. In effect, with the support of the US government, the US agribusinesses were introduced in a number of food-needy countries under the cover of promoting crop science and modern agricultural techniques.

The apparent success of the Green Revolution in Mexico enabled Rockefeller to use these technologies and seeds to spread to other countries worldwide in the 1950s and 1960s. The United States, for instance, had imported about half of its wheat needs in the 1940s but after using Green Revolution technologies after the Mexican experiments it became wheat self-sufficient in the 1950s and an exporter of the same crop by the 1960s. Encouraged by these results, the Rockefeller Foundation and the Ford Foundation, as well as many government agencies around the world, funded increased research in the new approaches.

India was chosen as an area to focus on after Mexico in the early 1960s because it was said to be on the brink of mass famine due to its rapidly growing population. Already in 1959 a team led by the US Department of Agriculture had published the Ford Foundation's researched report on India entitled: *Report on India's Food Crisis and Steps to Meet It*, in which it stressed the need for technological change, including improved seeds and the use of chemical fertilisers and pesticides in small, already-irrigated pockets of the country. The report advocated a technicalist approach instead of advising fundamental changes in the socio-economic system of land redistribution from the semi-feudalistic landowning system then in place in India as the path for the future agricultural development. The Ford Foundation had even gone beyond doing research to active funding of the country's Intensive Agricultural Development Programme as a test to the recommended strategy. This initiative provided rich farmers in irrigated areas with subsidised inputs, generous credit and price incentives. The World Bank backed this initiative with generous loans, thus providing a fertile ground for the Green Revolution in India [Ibid: 131].

Soon thereafter, the Rockefeller-Ford Green Revolution strategy was adopted by the Indian Government, after much prodding by the World Bank. The adoption of this strategy diverted the Indian government from pursuing its land reform policy, which had aimed at tenancy reform and the abolition of usury. These policies were dropped 'never to return' [Ibid: 132]. Before this was done, the Ford Foundation and their employee, Dr Borlaug, had already implemented the research results that had been obtained in the Mexico centre, and developed a new variety of rice in India called IR8 that produced more grain per plant when grown with irrigation and fertilisers. The production was limited to 20 per cent of land in the irrigated North and Northwest of the country. The programme ignored the huge disparities of wealth between the large feudal landowners in such areas and the poor and landless peasants. Just like it did in Mexico, where all hybrids were planted in the rich, newly-irrigated farm areas of the Northeast, in India the programme was promoted in pockets tied to large export agribusiness giants such as Cargill, while the regions with the vast majority of poor peasants remained poor and isolated.

The strategy behind the Green Revolution programmes was that hunger was to be overcome by using high-yield varieties, which would double yields on land by the application of new farm technology based on modern dam irrigation systems, genetically-engineered 'miracle seeds' and modern chemical fertilisers and pesticides. In a way, it was a 'transfer of technology' paradigm that was being advocated because it was argued that the existing traditional agriculture methods being practised in most undeveloped parts of the world were backward and incapable of innovation.

The results of the Green Revolution in India were indeed impressive. The initiative revolutionised the productivity of wheat and rice and the total grain production in the country by quadrupling their production in almost 40 years since its introduction, compared to the pre-Green Revolution period. This constituted a surge in production volumes from 48.1 million tonnes of rice and wheat in 1951-52, to 206.4 million tonnes in 2004-05.

But other estimates gave rather moderate performance records. According to a report produced by the Asian Development Bank as early as 1976, the growth rate in rice yields between 1963-67 and 1971-75 was a moderate 1.5 per cent per annum for South and Southeast Asia as a whole and in irrigated or wet (rice) fields. When it came to dry lands there was no evidence of major breakthroughs and growth rates were below one per cent for several countries [Pingali and Rosegrant, 1994].

However, these impressive and moderate records should not be attributed to the high-yield varieties (HYVs) and chemical use alone. There were other factors that contributed to this achievement. These additional factors were: first, the increase in the area of land that was extended to the production of these crops at the cost of the natural pulses and oil seeds. It is estimated that the area expansion alone contributed to the growth of one-third of the Asian rice output in the 1960s and one-fifth in the 1970s; second, the introduction of irrigation enabled production to take place two to three times a year unlike the situation in the previous period. This enabled production to keep pace with the increased population for almost three decades; third, Green Revolution famers were availed access to credit as well as the easy availability of the markets due to state price support – a fact that later became a problem in the rising indebtedness of farmers leading to suicides; and fourth, there were subsidies extended for farm inputs and the development of new varieties as well as the improved infrastructure to those areas producing rice and wheat for easy transportation of the crop and inputs.

Despite these achievements, the economic gains from the Revolution began to be questioned. This is because the cost of seeds, fertilisers and pesticides began to rise as economic profits also began to dwindle. At the same time, ecological costs in terms of loss of agricultural diversity, depletion of the soil content and the adverse impact caused by large dams proved damaging to the ecosystem. Finally, by benefitting rich farmers mainly, rather than the ordinary farmer, the new methods and seeds became a threat to the cohesion of rural society especially, leading to a widening rural-urban divide. The poor farmers who were impoverished due to onerous interest charged on loans were dispossessed and were made to drift to the urban areas where they became the urban poor.

Furthermore, there was a growing questioning of the technical achievements of the new agriculture, which ignored the socio-cultural impacts. Even in areas such as the Punjab, where the revolution was initiated, diseases of different kinds increased, which were caused by underground water pollution due to the increased use of fertilisers, especially nitrates and pesticides. Punjab, which was known as the food basket of India, was now referred to as the 'cancer state'. There was also a noticeable increase in suicides, which were committed among poor farmers due to indebtedness. It is estimated that over 125 000 farmers committed suicide in a period of 15 years and in one area alone. A scholar has estimated that some 8500 farmers from six cotton-growing districts committed suicide in a period of three years [Dharmitra, 2009: 8].

Other socio-economic adverse consequences include accidents and deaths of agricultural workers and farmers attributed to the lack of knowledge on the safe use of chemicals, which went unreported. In many of these cases poisoning occurred during spraying, mixing and diluting of pesticides, or due to the use of malfunctioning or defective equipment supplied by the chemical industries. Many of these workers and farmers were not trained in the use of these inputs and equipments, nor were they able to read instructions on the equipment or the manuals.

The real adverse impact was the total result of the achievements of the Revolution. The alleged 'increases' in farm yields began to decline due to growing soil depletion and the increasing use of fertilisers to revive them. This demonstrated the limits of the chemical inputs and the whole model of transfers of technology. It also raised fundamental questions about the role of scientific methods and knowledge systems connected with industrialised agriculture. The end product of the system was now blamed for the resulting global climatic change. Indeed, the issue of climatic change raised the question as to whether industrial capitalist agriculture was any more viable as a form of economy.

It became clear that high-yield seeds, chemical fertilisers and irrigation methods had played a large role in the Green Revolution and this forever changed the ecosystems and ecological conditions where various crops could be grown. For instance, before the Green Revolution, agriculture was restricted to areas with a significant amount of rainfall. But with irrigation, it was possible to store water, which could then be sent to drier areas, putting more land into agricultural production. This increased nationwide cultivation of crops and agricultural yields. In addition, the development of high-yield varieties meant that only a few species could be grown but this led to the emergence of monocropping and monocultures of single crops such as rice being grown. But by the same token, monoculture contributed greatly to the depletion of the soils and damage to the ecosystem. It became a vicious cycle, leading to the undermining of crop biodiversity.

In the Indian Punjab province, for instance, there were about 30 000 rice varieties prior to the Green Revolution. But with monocropping and monocultures, there were only around ten varieties left – all the most productive types. By promoting crop homogeneity, the remaining varieties were subjected to increased disease and pest attacks because there were not enough varieties to fight them off. In order to protect the few varieties left, it became necessary to increase the use of pesticides, chemicals and herbicides, which increasingly worsened the situation into a vicious cycle. Thus, along with the benefits gained from the new agriculture there have emerged several criticisms regarding this form of farming. Despite these criticisms, the revolution has achieved one success. It had created a large new market for US and foreign agribusiness multinational firms in developing countries where the Green Revolution had been introduced. What appeared as adverse impacts cause by chemicals, petroleum, machinery and other inputs were the achievements of agribusiness in the form of super-profits for those products. Moreover, the Green Revolution had opened up new avenues in advancing industrial agriculture in what came to be called the 'Gene Revolution' to which it immediately led [op. cit. 132].

[B] FROM GREEN REVOLUTION TO GENE REVOLUTION

The Gene Revolution and agribusiness should be seen as the main beneficiaries of the Green Revolution, and therefore an intensification of the drive to privatise the control of the seed as a way of controlling world food production, world population and world markets by US agribusiness. Indeed, the drive towards the Gene Revolution was prompted by the economics of the Green Revolution which had revealed that the proliferation of the new hybrid seeds in developing markets had reached a stage of declining returns, as we have seen above. This is because the high-yielding hybrid seed behind the Green Revolution had an inbuilt protection against multiplication and therefore it lacked reproductive capacity. Unlike the natural open-pollinated species whose seeds gave yields similar to its parents, the yield from the hybrid seeds was significantly lower than that of the first generation.

This meant that with the hybrid seeds the farmers had to buy new seed every year in order to obtain high yields if they applied the required chemicals. But at the same time, the lower yield of the second generation tended to eliminate trade for agribusiness due to the seeds being produced by middlemen without the agribusiness breeder's authorisation. In that situation, the agribusiness seed breeder tried to prevent the redistribution of the commercial crop seed by the middlemen. If the large multinational seed breeding companies could control the parental seed lines in-house, no middleman competitor or farmer would be able to produce the hybrid seed. This is the economics that led the global concentration of hybrid seed patents into a handful of giant seed agribusiness corporations; and it is this development that prompted a few of them, such as Monsanto, to lay the ground for the GMO seed revolution.

But it was the neo-liberal policy turn promoted by the pro-agribusiness regimes of Margaret Thatcher in the United Kingdom and the Ronald Reagan government in the United States that gave real muscle to the drive by removing the government regulation of the agribusiness research on genetically-modified organisms. The genetic engineering research field, which had developed a few years before out of the DNA (Deoxyribonucleic Acid) and RNA (Ribonucleic Acid) research activity, formed the basis of the new genetic research. At the same time, there had also emerged a 'science' of eugenics, which the Ford Foundation had promoted back in the 1930s, which was aimed at 'improving the quality of human species' but whose main objective was to control population growth by eliminating those human 'species' that were considered 'inferior'. This was the origin within which genetics emerged as a 'science'.

i. Biology and Eugenics – a reductionist manipulative 'science'

In that early period, the Rockefeller Foundation had funded the work of Margaret Sanger's Planned Parenthood Federation of America, which was initially known as the American Population Control League. Eugenics itself was a pseudo-science, which was first coined by Chares Darwin's cousin, Francis Galton. As it developed, Galton came to be referred to as the father of eugenics. His work was founded on Darwin's 1859 book, *On the Origin of Species*, in which Darwin had metaphorically applied the theories of Thomas Malthus on population to the entire field of vegetable and animal kingdom. As can be remembered, Malthus, who had repudiated these theories by the time of his death, had argued in his 1798 tract, *Essay on the Principles of Population*, that 'population tends to expand geometrically while food supply (tends) to grow only arithmetically, leading to periodic famine and death to eliminate the "surplus" population'.

In a paper read before the Sociological Society at a meeting in the School of Economies at London University on May 16 1904, Galton defined 'eugenics' as the science which deals with all influences that can 'improve the inborn qualities of a race'; also with those that 'develop them' to the utmost advantage. He posed the question: 'What is meant by improvement?' and 'what by the syllable *eu* in "eugenics"', whose English equivalent is "good"? There is considerable difference, he argued, between goodness in the several qualities and in that of the character as a whole. The character depends largely on the proportion between qualities, whose balance may be much influenced by education. He pointed out that we must therefore leave morals as far as possible out of the discussion and not entangle ourselves with the almost hopeless difficulties they raise as to whether a character, as a whole, is good or bad. Moreover, the goodness or badness of character is not absolute, but relative to the current form of civilisation [Galton, 1904]. It was with this definition of eugenics that Margaret Sanger, funded by the Rockefeller Foundation, embarked on her population control activities.

From this understanding, Galton had devoted most of the rest of his life exploring variation in human populations and its implications, at which Darwin had only hinted. In doing so, he eventually established a research programme, which embraced many aspects of human variation, from mental characteristics to height; from facial images to fingerprint patterns. This required inventing novel measures of traits, devising large-scale collection of data using those measures, and in the end, he discovered new statistical techniques for describing and understanding the data. Galton was interested at first in the question of whether human ability was hereditary, and proposed to count the number of the relatives of various degrees of eminent men. If the qualities were hereditary, he reasoned by analogy, that there should be more eminent men among the relatives than among the general population.

He obtained his data from various biographical sources and compared the results that he tabulated in various ways. He showed, among other things, that the numbers of eminent relatives dropped off when going from the first degree to the second degree relatives; and from the second degree to the third. He took this as evidence of the inheritance abilities. He also proposed adoption studies, including trans-racial adoption studies, to separate the effects of heredity and environment. From this he believed that a scheme of 'marks' for family merit should be defined; and early marriage between families of high rank should be encouraged by provision of monetary incentives. He pointed out some of the tendencies in British society, such as the late marriages of eminent people and the paucity of their children, which he thought were dysgenic. He advocated encouraging eugenic marriages by supplying able couples with incentives to have children.

But despite the fact that Malthus had repudiated his theories, the Rockefeller Foundation applied them as an ideological foundation and not as a science, which eugenics claimed to be pursuing, together with the Carnegie Foundation and other wealthy American family corporations that had 'embraced' his theories, which came to be called 'social Darwinism' to justify the accumulation of their vast fortunes. They did this on the ideological argument that the fortunes they were making were 'a divine proof of their superior species' survival traits over less fortunate mortals' [Engdahl, op. cit. 76]. This argument was also used by these rich families, especially the Rockefellers, for economic reasons: to refuse to pay taxes to the State on the ground that their taxes would be used for the benefit of the poor, preferring to keep the fortunes in the form of Foundations for 'philanthropic' purposes of their choice [Ibid.]. Eugenics was, therefore, used as Margaret Sanger put it, as a 'qualitative factor over the quantitative factor ... in dealing with the great masses of humanity', in which her Population Control League – a racist association which advocated sterilisation of the inferior – was part of her drive to control population. This is what the Rockefeller Foundation funded and supported in the name of promoting the 'science' of eugenics, which later turned deadly against nature.

The 'science' of eugenics was used by the Nazis to exterminate Jews and other 'inferior' races. Beginning with the idea of a blue-eyed Nordic 'superior' race, which was not a mere fantasy of the Nazis, the Rockefeller Foundation in their research, led to a book, *Blood of a Nation*, by David Starr Jordan of Stanford University, which argued that poverty was not an economic or social fact but an inherited trait in the same way that talent was inherited. The Carnegie Institute went on to establish a Eugenics Record Office on Long Island, outside New York City, where millions of index cards were compiled of the bloodlines of millions of ordinary Americans who were deemed to be 'inferior' for purposes of their 'removal' from society [Ibid: 77]. The ideals for the superior type were tallness, blond, blue-eyed Nordic types.

The Rockefeller Foundation poured millions of dollars into various eugenics and population projects as part of its work. A branch of eugenics called Applied Eugenics, proposed by one, Dr. Paul Bowman Popenoe, a US Army venereal disease specialist from World War I, identified the numbers of millions of Americans who would end up in mental hospitals. Some five million Americans who were regarded as 'deficient intellectually' with less than '70 per cent intelligence' were categorised as constituting a liability rather than an asset to their race. This categorisation addressed to an elite audience constituted what the eugenic movement called 'negative eugenics' – the systematic elimination of 'inferior' beings, whether mentally inferior, physically handicapped or racially non-white' [Ibid: 78].

Taking this American system, the Nazis developed a programme in Germany for the sterilisation of Jews, Gypsies and other 'defectives' by force. In 1933, based on these experiments, they passed a Sterilisation Law, which they described as an American Model law. The law was adopted in July 1933 and signed into law by Hitler himself for immediate implementation. Under the law some 400 000 Germans were diagnosed to be manic-depressive or schizophrenic and forcibly sterilised and thousands of children so categorised were simply killed. After the war and the disgrace of the Nazi regime, the remnants of the eugenics team of 'scientists' camflouged themselves under a new 'science' called Genetics under which they claimed to be founding a new science for humanity. To be sure, its first president was Hermann Josef Muller, who was a Rockefeller Fellow and worked at the Kaiser Wilhelm Institute for Brain Research in 1932 under the Nazis.

By the 1950s, genetics had attained a respectable status in the academy in the United States. By that time the large agribusiness corporations had taken over small farms and, applying genetics as a science, were producing large numbers of cattle, pigs, chickens and other food products. The use of chemicals and antibiotics to treat cattle assembled in factories began to produce epidemics and incidences; and diseases of different kinds such as obesity, allergies, salmonella poisoning and e-coli, all became everyday events. By 1998 a new university-wide research programme headed by one, Ray Goldberg, was set up to examine how the new genetic revolution 'would affect the global food system' [Ibid: 146]. The idea was to integrate the Gene Revolution into an agribusiness revolution as the next phase to the Green Revolution, in which the small farmer was to become a 'tiny player in the giant global chain'.

The new agribusiness was to comprise entirely new sectors created by the latest developments in genetic engineering, including the GMO creation of pharmaceutical drugs from genetically-engineered plants, which Goldberg called 'the agri-ceutical system'. From this Goldberg observed that the genetic revolution 'is leading to an industrial convergence of food, health, medicine, fibre, and energy business' [Ibid: 146]. Engdahl concludes that from the Green Revolution to the Gene Revolution, the Rockefeller Foundation was in the centre of developing the strategy and means for transforming how the planet would feed itself in the future [Ibid: 147]. By this time, agribusiness was ready to capture the food market as a global 'golden rice bowl'.

The ideological argument for the new move was that the world population was growing at a fast rate and that there was a need to address the needs of food security for the world. To achieve this, the Rockefeller Foundation between 1982 and 1990 spent millions of dollars on 'molecular biology', which was to engage in plant breeding. Molecular biology was emerging from the science of genetics. In December 1984, the Foundation adopted a 10-15 year research programme to apply the new 'science' and techniques to the breeding of rice, which was the dietary staple food product of the majority of the planet's population, in order to control its supply and market. So the issue was not feeding the world population but using the new molecular biology to control world food markets in a key staple food that the world population relied on, and in doing so eliminate the natural seeds that supplied the population. The Foundation had also recruited chemists and physicists 'to foster the invention of a new science discipline', which it named 'molecular biology' to distinguish it from classical biology. This renaming, according to Engdahl, was done in order 'to deflect any blunt growing social criticism of its racist eugenics' because the Nazi regime had given the name eugenics 'a bad name' [Engdahl, op. cit. 153].

Borrowing generously from its work in race genetics, the Foundation scientists, led by Warren Weaver, the Foundation's first director of the natural sciences division, developed the idea of molecular biology 'from the fundamental assumption that almost all human problems could be "solved" by genetic and chemical manipulations'. This was another example of the use of metaphor and analogy to create a 'science', as we have seen above. The term 'molecular biology' was then invented by the Foundation to 'describe their support for research to "apply techniques of symbolic logic" and other scientific disciplines to make the "biology" they were developing "more scientific". This is because the scientists working for the Foundation had concluded from the earlier experiments they had carried out under eugenics that echinoderm larvae could be chemically stimulated to develop in the absence of fertilisation. From this they had drawn the conclusion that 'the science would eventually come to control the fundamental process of biology'. They further came to the conclusion that this was 'the ultimate means of social control and social engineering, which they had already called eugenics' [Ibid: 154].

But, according to Fuller, this 'turned out to be much narrower in focus and less interventionist in spirit' than the Foundation wished [Fuller, op. cit. 142]. Weaver, on his part, had come to the conclusion that in order to move forward in a more interventionist manner, the Foundation would have to 'invade the biological and medical sciences' to achieve their objectives. In this respect, they considered the 1930s to be a more amiable period for such 'a friendly invasion by the physical sciences' into this area. The methodological approach for the 'friendly invasion' involved the application of *reductionism*, which held that all living creatures, according to Rene Descartes, were 'machines whose only use was their genetic replication'. Although this theory had been discredited, the Rockefeller molecular biologists still applied it to mean that a complex life form *could be reduced* to a basic building block or its '"elementary seed", from which all traits of the life form could be deducted'. Once the scientists made this reductionist decision that all organisms were 'reducible' to genes, they could conclude further by the same reductionist method that 'organisms had no inherent nature. Anything was fair game' [Engdahl, op. cit. 155].

This crude, outmoded view of organisms became embedded in genetic determinism, an extreme version of the classical genetics that dominated biology from roughly the 1930s to the 1970s, when genetic engineering took over and continues to do so in social and public policy. Since prehistoric times, it was believed that all living things inherited traits from their parents and this understanding has been used to improve crop plants and animals through selective breeding. However, with the advent of the modern genetics reductionist 'science' that began with the work of Gregor Mendel in the mid- nineteenth-century, this understanding began to change. Although Mendel did not know the physical basis for heredity, he nevertheless observed that organisms inherit traits via *discrete* units of inheritance, which are now called genes. According to the new theory, genes correspond to regions within DNA, a molecule composed of a chain of four different types of nucleotides, which constitute the sequence of the genetic information, which concrete organisms are supposed to inherit.

The sequence of nucleotides in a gene is supposed to be translated by cells to produce a chain of amino acids creating proteins – the order of amino acids in a protein corresponding to the order of nucleotides in the gene. This relationship between nucleotide sequence and amino acid sequence is known as the *genetic code*. The amino acids in protein determine how the gene folds into a three-dimensional shape and this structure is, in turn, responsible for the protein's functioning. In this reductionist structure, proteins carry out almost all the functions needed for cells to live. A change to the DNA in a gene can change a protein's amino acids and thereby change its shape and function. According to the theory, this sequence can have a dramatic effect in the cell and on the organism as a whole. Although genetics plays a large role in the appearance and behavior of organisms, it is the combination of genetics with what an organism experiences that determines the ultimate outcome. For example, while genes play a role in determining an organism's size, the nutrition and health it experiences after inception also have a large effect.

But no gene ever works in isolation, to be manipulated to serve any purpose: rather, it operates in an extremely complicated gene network. The function of each gene is dependent on the context of all the other genes in the genome. For this reason, the same gene will have different effects from individual to individual, because other genes are different. There is so much diversity within the human, animal, and plant populations that each animal, plant or individual is genetically unique. If the gene is transferred to another species, the transference is most likely to lead to unpredictable results because of the very complicated ecology consisting of the interconnected levels of the genome, the physiology of the organism and its external environment. Putting a new gene into another organism is bound to create disturbances that can propagate out to the external environment. Conversely, changes in the environment will be transmitted inwards and may alter the genes themselves. Hence, genetic engineering profoundly disturbs the ecology of genes at all levels, and this is where the problem and dangers of genetic engineering arise [Mae-Wan Ho, Harmut Meyer and Joe Cummings, 1998: 146-153].

But by a reductionist metaphor, the Rockefeller scientists argued by analogy that 'genetic programming' could be used for their purposes, although no scientist was able to generate an organism from a computer genetic programme to resemble the natural organism. This is because one needed to know more than the gene products in order to explain the emergence of shape and form of organisms. But the objective of these scientists was to find out how plant and life forms could be manipulated in a Rockefeller genetic model aimed at the control of plant and life forms in order to influence social policy. The idea was to use 'molecular biology' to map the structure of the gene, and to use that information obtained to influence social and public policy. The Foundation's objective was 'to correct social and moral problems including crime, poverty, hunger and political instability'. According to Philip Regal:

'From the perspective of a theory of reductionism, it was logical that social problems would reduce to simple biological problems that could be corrected through chemical manipulations of the soil, brains, and genes ... The Rockefeller Foundation used its funds and considerable social, political, and economic connections to promote the idea that society should wait for scientific inventions to solve its problems, and that tampering with the economic and political systems would not be necessary. Patience, and more investment in reductionist research, would bring trouble-free solutions to social and economic problems ... The project was in the general spirit of Bacon's New Atlantis and Enlightenment visions of a trouble-free society based on the mastery of nature's laws and scientific and technological progress [Quoted in Engdahl, op. cit. 157].'

Therefore to achieve this *combination* of control, the heart of the genetic engineering of plants, unlike the earlier long-standing methods of creating plant hybrids by cross-breeding two varieties of the same plant to produce a new variety with specific traits, involved *introducing foreign DNA into a given plant*. The combining of genes from different organisms was termed recombinant DNA or rDNA. This gene transformation required a *tissue culture* or the regeneration of an intact plant from a single cell that had been treated with hormones or antibiotics and forced to undergo abnormal development. Therefore, entirely new genes were made in a laboratory and inserted into the genomes or natural organisms to make *genetically-modified organisms*. But the process was not 'neat', 'scientific' or 'reliable' as the molecular scientists claimed. According to a biologist, Dr. Mae-Wan Ho, the Head of the London Institute of Science in Society, the process was 'uncontrollable and unreliable, and typically ends up damaging and scrambling the host genome, with entirely unpredictable consequences' [Ibid: 159].

Thus a reductionist imitative 'science', that disregarded the real and holistic science that took into account all the implications of such pseudo-scientific techniques, was being applied by analogy, and metaphorically, to promote and destroy both plant and other life forms by the Rockefeller Foundation for the selfish purposes of making super-profits and an attempt to control the world in the name of 'progress'. These scientists did not examine or seriously consider such risks to human life and the plant world because this was supposed to be applied to 'inferior' groups of people. With this haste, by 1973 the first genes had been spliced into existence and from then onwards, the recombinant methods were spread to many parts of the world without controls in their application. This was made easier because the Reagan and Thatcher administrations, through their neo-liberal global policies of deregulation and privatisation, were able to argue that the products of such 'science' had to be accepted as 'intellectual property' of the agribusinesses under the new World Trade Organisation rules called Trade

Related Intellectual Property (TRIPS), under which life and plant forms genetically-modified in this way could be patented as 'private property' and protected under the international legal regime. Thus, through this manipulative reductionist 'science' a few monopolies were able to establish a 'legal' and 'moral' basis for controlling not only the basis of livelihood and population reproduction, they had gone further to claim the right to reproduce or destroy real life forms and plants.

ii. Control of food markets and populations

The mapping of the rice genome was achieved in 1984 as one of the channels for exercising food control. In order to achieve its objective of 'fighting world hunger' for super-profits, the Rockefeller Foundation in 1987 made grants of $5 million to set up 46 science laboratories throughout the world for mapping the rice genome and training a network of international scientists at both post-graduate and doctoral levels to master and advocate the world-view application of genetic engineering and 'molecular biology' in particular. They tried to develop an elite fraternity of scientists to cultivate a sense of mission and belonging, believing in the 'betterment of man' through the manipulative 'scientific' applications. They claimed to have 'discovered' a vitamin A which was to be inserted into the rice to solve the 'Vitamin A deficiency' in undernourished children in the developing world, knowing full well that this was a staple that had been developed over 12 000 years and fed over 80 per cent of the population in countries such as China, India and West Africa. Now the Foundation wanted to have a monopoly control over this crop by contriving genetic engineering in order to patent it as its private property.

Already under the Green Revolution rice had been 'improved' by using High-Yield Varieties. This episode in Asia did not produce any sustainable results; only short-term increases in production. The strategy had succeeded only in drawing the Asian peasants into the vortex of the global trade system in which they were not able to compete. Moreover, the market for fertilisers, high-yield seeds, pesticides, mechanisation, irrigation, credit systems and global marketing schemes packaged by agribusiness monopolies had the objective of replacing small-scale farming with the new agribusiness enterprises. The earlier Green Revolution had used the Philippine-based International Rice Research Institute (IRRI) to create a gene bank comprised of one-fifth of the world rice varieties of the region.

The emerging effect of this Green Revolution strategy was the disappearance of these varieties. Instead of the banks being the reservoirs for these seed varieties, they became prison camps for their elimination and disappearance.

The gene banks now became the vehicle for the proliferation of the Rockefeller GMO rice seeds and, as the industry gained a foothold, new institutions were created to capture even more crop varieties for the GMO agribusinesses to illegally modify them genetically and render them private property for agribusiness. Agribusiness was able to insist that the newly genetically-modified seeds had themselves to be 'protected' by patents held by them under the World Trade Organisation (WTO) TRIPS agreements. These monopolistic practises, which were imposed on the world by the US government in their negotiations during the Uruguayan Round leading to the WTO, strengthened agribusiness in using the gene revolution as the next giant step in consolidating their grip on the global food supply control through the new World Trade Organisation.

The step to develop genetically-modified crops was first tried in Argentina where a new 'Peronist' but neo-liberal regime under President Carlos Menem had taken control in 1989. Here Monsanto had taken power by working on the Argentinean government to accept experimentation with the genetically-modified soybean seeds. This was hailed as the 'Second Green Revolution'. Within a short period of eight years, Monsanto had managed to promote a worldwide acreage of GMO crops of 167 million acres by 2004, an increase of 40-fold. This represented a 25 per cent use of the total agricultural cultivated land in the world, which already 'suggested that GMO crops were well on the way to fully dominating world crop production, at least in basic crops, within a decade or even less' [Engdahl, op. cit. 176-7].

President Menem had argued that in order to get rid of a debt-saddled economy, Argentina must accept a monoculture of soybean production in order to earn sufficient foreign exchange to be able to pay its debts. But little did Menem understand that those who dominated agribusiness also dominated the financial system that had created the debt that had weighed down Argentina's banks in the first place. These banks were the Rockefeller JP Morgan Chase Bank and others. The drive to transform Argentinean agriculture into GMO soybean production turned the Argentinean population into what Engdahl has called 'a human guinea pig of the project'. Even before the growing of the crop had begun in the US,

Argentina had become a secret experimental laboratory for developing GMO crops. The genetically-modified soybean had a gene from the bacterium – a strain CP4 – inserted into its genome by a gene gun and when it was sprayed by a non-selective herbicide called glyphosate, the active ingredient had the capacity to kill conventional soybeans grown by peasant farmers adjacent to the Monsanto Roundup Ready crops in the fields due to wind-borne contamination.

The spread of the soybean plantation was promoted by a 'labour-free' system called 'direct drilling', which was pioneered in the USA for purposes of 'saving time and money'. This led to huge stretches of land in which large mechanised equipment could operate around the clock without stopping. This mammoth machine was able to dig holes into the ground and to automatically plant the GM soybean seed several centimetres deep and then press soil on top of it. With this machine, just one person could plant thousands of acres of land, thereby saving both time and money for agribusiness. But after a few years of such planting, the herbicides sprayed to kill the weeds began to show special intolerance to glyphosate, which required ever stronger doses of the chemical herbicides to cope with the planting of the soybeans.

Through these means, the Argentinean economy was completely transformed in less than a decade into a monoculture. The impact was too great. The small farmers were dispossessed and destroyed as well as their soybean varieties. By the year 2000, after only four years of adopting Monsanto soybeans and mass production techniques, over 10 million GMO soybean hectares had been planted. By 2004 the acreage had expanded to 14 million hectares, which resulted in the clearing of more forests, as well as occupying the traditional lands that belonged to the indigenous peoples of the region, to create more room for soybean production. Argentina was well-known for its beef industry, but within a few years, the large dairy and beef herds which had roamed freely on large grasslands, were forced into cramped mass cattle feedlots to make way for the lucrative soybean GM business. Fields of traditional legumes, cereals, lentils, peas and green beans had all vanished.

By 2004, almost half of all the agricultural land in Argentina was taken up for GM soybean production, and between 90 per cent and 97 per cent of these were Monsanto products. In the same period, over one-half of the dairy farms had been reduced, and milk production had declined, so that Argentina had to import milk to meet its own needs, whereas before it was an exporter of milk. As thousands of farmers and workers were forced out of production, poverty and malnutrition intensified. By 1998, Argentina, which had enjoyed one of the highest standards of living in Latin America with a population living in poverty below five per cent, had the people living in poverty escalating to 30 per cent, and by 2002, the people in poverty had gone up further to 51 per cent. Malnutrition levels rose to around 11 per cent and 17 per cent of the total population of 37 million.

In the meantime, in Brazil, the country under President Lula da Silva, had also been forced to throw in the towel by submitting to Monsanto pressures to permit GM soybean cultivation in that country. Here, too, production increased very quickly so that by 2006, both Argentina and Brazil were accounting for more than 81 per cent of all world soybean production, 'thereby ensuring that practically every animal in the world feeding on soymeal was eating genetically-modified soybeans' including hamburgers mixed with soybeans [Ibid: 190].

But far from providing conditions for fighting hunger, Argentina began to feel a shortage of food, especially after 2002, and was faced with the destruction of its traditional agricultural and cattle production. Fearing food riots, Monsanto and giant soybean users such as Cargill, Nestle, and Kraft Foods responded by giving out free soybean-based food packets to the population. The food packets they gave out were cynically being used as an advertisement intended to increase their markets for the crops. Much of this 'food' was in fact soybeans intended for animal consumption. And unknowingly to Argentineans, the raw and processed soybeans contained a series of toxic substances which could cause cancer. Swedish studies had shown that one of these substances, called Trypsin, was linked to stomach cancer. Thus, far from 'fighting hunger', Monsanto and other GM agribusinesses had created poverty and hunger apart from destroying the agricultural base of countries.

The GMO propagation has also contributed to the destruction of biodiversity, especially where these technologies are being imposed on poor farmers by biotech conglomerates taking advantage of the hunger of these poor communities. In Ethiopia, for instance, the farmers were seduced with kits of GMO seeds under the pretence that the kit was a 'gift' and 'food aid' intended for the 'rehabilitation' of agricultural production in the wake of a major drought that struck the country. The seeds were planted, but the farmers soon found out that they could not replant the seeds without paying royalties to the GMO conglomerate, Monsanto. The farmers also discovered that the seeds could only be replanted if they used the chemical inputs which included fertilisers, insecticides, and herbicides produced and distributed by the same agribusiness corporations, from which they now had to buy at great cost.

Scientific or industrial agriculture, of which the GMO is the latest fad, is altering the entire peasant agricultural economy as the peasants have found themselves gripped into the web of the GMO agribusiness conglomerates' entrapments. The farmers have also found that with the widespread adoption of GMO seeds, a major transition is occurring in the structure and history of settled circular agriculture since its inception 12 000 years ago. Michael Chossudovsky has noted these consequences, which is spelling the end of agriculture as we have known it:

> 'The reproduction of seeds at the village level in local nurseries has been disrupted by the use of genetically-modified seeds. The agricultural cycle, which enables farmers to store their organic seeds and plant them to reap the next harvest, has been broken. This destruction pattern – invariably resulting in famine – is replicated in country after country, leading to the worldwide demise of the peasant economy [Chossudovsky, 2010: 164].'

This, the peasant plight, is not a short-lived experience: it is a worldwide destruction of agriculture from its original epistemological foundations that were expressed in the myth of Osiris and Isis, in which plant life was associated with the birth and death of the crops in the cycle of a never-ending reproduction, resulting in the agricultural economy before the advent of modern industrial agriculture. According to this epistemology, the seed was born with every season of planting in a circular pattern of yearly reproduction. The seed 'died' when the crop was harvested and after it had produced enough food for humans who planted it. The new seed from the crop was then stored and replanted in the next season when it was 'born' once again to sustain life and the ecosystem.

This was the circular pattern of agriculture which sustained life and in which nature also renewed its cycle. This epistemology of agriculture has slowly been coming to an end through the introduction of 'scientific agriculture', which has relied on high-yield crop varieties, chemicals and mechanical technology that has damaged the soil and impacted negatively on the ecosystem. The latest phase of this form of agriculture in the form of the GMO seed brings this historical agriculture to an end by destroying the basis of the reproduction of the seed without a large amount of chemicals and other non-organic inputs. These inputs have killed the soil in many places where the Green Revolution has been introduced and where the GMO agribusiness has now come in to finish the job of destruction of the natural seed, which is the inheritance of the human race from nature. This destruction has the implication of bringing to an end the biodiversity on which life depends and therefore human life itself.

The introduction of the 'terminator seed' by Monsanto will add the last death-blow to the natural seed, if it has not already done so. On October 4 2011, Monsanto took the first step to 'commercialise' the 'terminator seed' genetics. The Ford Foundation was not too far behind these attempts, for the genetics revolution was the brainchild of the Foundation from the very beginning. It had spent over $100 million for the advancement of research in this endeavour. But later the Foundation tried to distance itself from the step to commercialise the terminator when there was a great deal of public concern about it. They twisted their support to mean that the Foundation, instead, wanted plant biotechnology to reach the poor farmers in the developing countries, when in fact the very objective of the terminator seed was part of its own design to terminate any capability they still possessed to use the natural seed in competition to the genetically-modified seed.

The objective of developing the terminator seed by Monsanto with the support of the Ford Foundation was indeed to prevent the germination of harvested grains as seeds. The technology was intended to block farmers from saving seeds for re-sowing. The aim was to produce a new technology 'which would allow them to sell seed that would not reproduce' [Engdahl, op. cit: 257]. The new technology was named: Genetic Use Restriction Technology (GURT) which popularly came to be called the terminator seeds. It was developed to 'protect corporations from unscrupulous farmers who might try to re-use patented seed without paying'. In short, the idea was to deny the right of farmers to use the natural seeds they had developed and preserved for centuries and substitute a monopoly over seed by a few agribusiness entities for private profit of the owners. As Engdahl has observed:

'The terminator looked like an answer to the agribusiness dream of controlling world food production. No longer would they need to hire expensive detectives to spy on whether farmers were re-using Monsanto seeds [Ibid: 258].'

To ensure the GURT strategy would be implemented without any hitch, the terminator corn, soybean and cotton seeds were genetically-engineered and programmed to self-destruct and 'commit suicide' after a single harvest season. The gene seeds had in-built toxins which were inserted before the seed ripened in such a way that at the appropriate time the 'plant embryo would self-destruct' [Ibid.]. Where a farmer succeeded in getting 'traitor seeds' from the 'illegal' seed market, it was made sure that such farmers would not get the special chemical compound needed to 'turn on' the plant's resistance gene. This assured the concerned agribusiness a virtual captive market for the sale of their products. Further, it turned out that it was cheaper to produce the 'traitor technology' than the complicated terminator seed because with the 'traitor' technology it was also possible to develop GMO plants that needed to be 'turned on' in order to grow or become fertile. The US government even went as far as supporting this form of technology as a 'Technology Protection System' for its agribusiness monopolies.

Due to public uproar and farmer protests throughout the world, Monsanto and the Ford Foundation tried to backtrack. They announced that they had decided not to commercialise the sterile terminator seed. But this was a shrewd tactic to deceive public opinion. Thus, while it called off the merger with another GMO promoter, Delta & Pine Land, in late 1999, the latter, with the US Department of Agriculture (USDA), continued with the programme to perfect Terminator and Traitor technologies. The USDA's commitment was that it wanted the technology to become available as widely as possible, 'so that benefits will accrue to all segments of society' [Ibid: 266]. In 2001, it signed a licence agreement with Pine & Delta allowing it to commercialise the terminator technology for Delta's cotton seeds. In 2003, Monsanto emboldened, began to repackage the technology as an 'ecological plus' and began to spread its use as a way of controlling the spread of GMO seeds by wind or pollination and the contamination of non-GMO crops.

Faced with the resistance to GMO crops from the EU and many African countries, the USDA used 'whatever means possible', including famine aid, to get those seeds to India and many African countries. In Asia, Monsanto had resorted to bribery to push its GMO seeds for the internal Asian market. The strategy of the USDA and Monsanto was to spread the seeds as rapidly as possible across the globe so that once they had been planted and taken root in some key countries, the crops would spread to other countries. In Latin America, the two partners tried in hothouse fashion to breed corn plants that could make 'anti-sperm antibodies' and a 'contraceptive corn' where they had taken some antibodies from women, isolated the genes that regulated the manufacture of those infertility antibodies, and, using genetic engineering techniques, inserted the genes into ordinary corn seeds used to produce corn plants.

They even went as far as taking another step from the Terminator and Suicide seeds to 'Spermicidal Corn'. This involved the manufacture of a corn variety which, when consumed, would make the human male sperm infertile. Such a spread could have been undertaken by companies that supplied cornflakes. This would be one way of promoting birth control and population limitation in certain countries. These genocidal imaginations were not very different from the eugenics projects in Nazi Germany, which the Rockefeller Foundation had supported as we have seen. Now the 'solution' would be found on the 'supply side', which would limit world population 'by going after the human reproductive process itself' [Ibid.]. Towards this end, the Rockefeller Foundation had funded the World Health Organisation (WHO) to embark on a 'reproductive health' programme, which would apply an 'innovative tetanus vaccine' which it had developed for this purpose. This was exposed in Mexico by Catholic lay organisation, Comite Pro Vida de Mexico.

In 2006 Monsanto finally exposed its real intentions when, on August 15, it decided to acquire the Delta and Pine Land now holding a technology regarded as 'a controversial genetic engineering technology that makes sterile seeds'. Rockefeller did not this time attempt to disassociate itself from the deal. This time the indigenous peoples of the Andean and Amazon regions made noise about the acquisition and its impact on their economies. They argued that Terminator threatened not only their seeds but also the knowledge and the invaluable experience they possessed about their crops, and hence their livelihoods, since their lives were built around seed saving and seed exchanges between plant breeders during barter markets. They argued:

'Terminator technology would have a concrete impact on the knowledge systems by jeopardising the availability of seeds for collective exchange and breeding. As a consequence of Terminator, the very processes of adaptive interaction between man and the climatically complex Andean and Amazonian ecosystems, which have allowed for the evolution and current vitality of a highly specialised body of indigenous knowledge, would be paralyzed [Ibid: 299].'

This deal between Monsanto and Delta closed the circle of Monsanto emerging as the ultimate monopolist of agricultural seeds for nearly every variety. In August 2005 it had lodged a patent for the patenting of the gene of a pig, which would have given it rights to collect licence fees for pig offspring produced by its method; and hence of intellectual property rights to particular farm animals and particular heirs of livestock. Having been empowered by a US Supreme Court decision in the case of Diamond vs. Chakrabarty [1980], which ruled that 'anything under the sun that is made by man is patentable', Monsanto tried to patent entire animal genetic seed lines of plants and animals for its benefit with the declared aim of controlling both oil and food markets throughout the world.

These developments sounded alarm bells even in the Vatican. This was after the EU was reported to be on the verge of granting Monsanto and Delta a patent for a terminator seed under the pressure of a US Trade Representative because the representative accused the EU of 'flouting WTO rules'. A number of EU member countries had declared their countries 'GM-free'. The Pope was reported to have urged these companies to 'stop playing God'. A secret diplomatic cable from Miguel Díaz, the US Ambassador to the Vatican, had revealed that the US regarded population growth as a major cause of climate change. The dispatch, released by the whistle-blower website, *Wikileaks*, criticised the Vatican for arguing that unsustainable lifestyles in developed countries – and not population growth worldwide – is to blame for global warming.

But Diaz was reported to have gone on to lament that 'the Vatican will continue to oppose aggressive population control measures to fight hunger or global warming'. But according to *Wikileaks* in January 2011 revelations: 'Recent conversations between the Holy See officials and USAID ... confirmed the cautious acceptance of biotech food by the Holy See'. This was revealed by Christopher Sandrolini, a US diplomat to the Holy See, in a cable dated August 26 2005. The cable revealed that the Vatican was not concerned about the safety, science and legitimacy of biotechnology and that, in fact, the mainstream opinion in the Vatican was that the GMO science was 'solid'. The Vatican did have an issue with the economic impact of GMOs on farmers because the Church was concerned that 'these technologies are going to make developing world farmers more dependent on others, and simply serve to enrich multi-national corporations'.

The US diplomats, however, actively lobbied the Vatican and tried to assuage the concerns by referring to competition among companies and the regulatory process in individual countries as safeguards against these concerns. Thus, even the Vatican has been subjected to pressure, and its policy on the issue is not clear. The only hope is with the poor farmers themselves in resisting this global monopoly onslaught on their livelihoods.

iii. Green and Gene Revolution in Africa?

African leaders have praised their Asian counterparts for having embarked on a new approach to the agricultural economy and judged their 'backwardness' to be the result of having failed to embark on the same path of 'economic development'. They have, from time to time, expressed their frustrations by declaring that they had 'missed the bus' of the Green Revolution. This 'failure' was later compounded by 'the lost decade' of the structural adjustment policies of the World Bank and the International Monetary Fund, which redirected agriculture from food self-sufficiency to cash crops for exports to pay accumulated 'sovereign' debts. As a result, African leaders have recently hastened to adopt Green Revolution strategies, which Asia and Latin America had adopted in the mid-1960s, in order to 'catch up' and boost food production and the agricultural economy in general promoted by the same old and new players.

The push to engage Africans in an 'African Green Revolution' is in fact intended to kill two birds with one stone because, unlike Asia and Latin America, which promoted Green Revolution strategies without other pressures, Africa is being required to embark on the Green Revolution while at the same time being advised to accept GMO seeds for a 'Genetic Revolution' or Gene Revolution. As we have seen, agro-dealers are being deployed all over the continent to spread 'improved seeds' – almost 'freely' – while at the same time driving the introduction of chemical fertilisers and pesticides, but with the objective of entrapping the farmers into the GMO-driven agribusiness economy. As we saw above, the famine in Ethiopia was used to give out 'free' food kits consisting partly of the GMO seeds for the peasants to plant on the pretext that this would get rid of famine in their country.

As a result, African governments have been prevailed upon by a number of agribusiness and philanthropic Foundations to embark on the Green

Revolution to save their continent. At their African Union ministerial meeting held in Windhoek, Namibia, to address 'African Agriculture in the 21st century: Meeting the Challenges, Making a Sustainable Green Revolution', African governments called for a 'uniquely African Green Revolution' to help boost agricultural productivity, food production and national and regional food security. They reiterated the urgent need for an African Green Revolution that does not depend only on improved seeds and fertilisers but is built on overall public investment in rural development, rural infrastructure, education, credit support, research and development, and technology development and dissemination. This sounds like a reflection on the Asian experience but it is not.

The inference here was that 'Sustainable Green Revolution in Africa', unlike that of Latin America and Asia, should be tailored to the highly diverse agro-ecological conditions on the continent as well as the diverse farming systems and socio-cultural contexts. But there was still the realisation that there was a need to reverse the spectre of poverty, hunger and famine that was being experienced in parts of the continent, and at the same time take into account the environmental degradation on the continent. From this it would appear that African governments who had at first been enthusiastic about the need for the Green Revolution had, by the time of the proposal by some 'philanthropic' organisations under the banner of the Alliance for the Green Revolution in Africa (AGRA), developed cold feet on the idea and were now expressing more caution on the idea.

Despite these cautions, which took into account the adverse consequences of the Green Revolution in Asia, the same kind of solutions that were recommended for Asian and Latin-American agriculture in the 1960s were nevertheless being pushed by the 'philanthropic' institutions for Africa. The same players that initiated the Green Revolution in Mexico and Asia – plus new ones – are the ones behind the Green Revolution Alliance for Africa, which has been touted since the 1990s. The drive was given a renewed impetus in 2007 when the Rockefeller and Bill & Melinda Gates Foundations launched the so-called Alliance for a Green Revolution in Africa (AGRA), to which former UN Secretary-General, Kofi Annan, gave support. Millions of dollars have been placed in the coffers of a host of carefully-selected individuals and companies to lay the ground work for the industrialisation of African agriculture and the creation of markets for agribusiness [Mayet, 2009: 12-13]. The role-players include corporations such as the Citizens Network for Foreign Affairs (CNFA) and the International Fertiliser Development Centre (IFDC), both of which are enmeshed with the corporate interests of giants such as DuPont Crop Protection and Monsanto within the AGRA projects in selected African countries.

The Bill & Melinda Gates Foundation is investing funds into biosafety projects that are part and parcel of the biotechnology industry, which focus on 'pie in the sky' nutritionally-enhanced, genetically-modified and 'biofortified', 'climate friendly' and drought-tolerant crops. According to Mariam Mayet 'This is done to win over the hearts and minds of reluctant Africans, while paving the path for the gene giants to gain a firmer and more respectable foothold in Africa' [Ibid: 12].' These funds are being used to usher in the two 'revolutions' in African agriculture at once and in tandem: one based on the classical Asian and Latin American Green Revolution models and the other based on the genetically-modified (GM) technologies, as we have seen. Thus, instead of promoting the sustainable model demanded by the African leaders in Windhoek, Namibia, the objective of the AGRA is to enshrine the dominant industrial agricultural model based on agro-exports, free trade, and the use of chemical-intensive, large-scale monocultures and GM organisms, as well as the transfer of the technology model.

The adoption of these belated imposed models is now being propagated and applied through the NEPAD (New Partnership for Africa's Development) via the Comprehensive African Agriculture Development Programme (CAADP) and the Framework for African Agricultural Productivity (FAAP). African leaders have thrown in their towel to give support to the programme – as usual without any reflection about its impact on their economies in the long-run. Their interests are short-term and they are not capable of taking a long-term view of policies advocated by multinational corporations and the 'donor', 'philanthropic' institutions. This long-term view is taken by the multinational corporations, which attests to the fact that Africa is still a neo-colonial enclave of states under the control of these corporate institutions.

Seductive promotion methods and techniques are being used by individuals employed to promote AGRA, such as the use of 'smart subsidies' to convince poor African farmers to use the Green Revolution and gene technologies. This is taking advantage of the situation in Africa where most governments do not have biosafety legislation in place; and where legislation for the protection of intellectual property rights of the farmers and rural communities is non-existent. Academic institutes such as Jeffrey Sach's Earth Institute of Columbia University have also signed a five-year agreement aimed at delivering 'the best science' policies and technologies to 'sustainably' improve agriculture for African small farmers. But Sachs is a great supporter of GM crops and therefore the only 'sustainable' science he can deliver in form of policy is to promote GMOs in Africa. He believes that the great promise GMOs present for small farmers is the fact that the 'the technology is delivered in the seed'.

The main focus of AGRA is on crop breeding, for which a five-year plan has been set to develop some 100 new seed varieties from core crop seeds such as maize, cassava, sorghum and millet. In effect, this means that Monsanto and other GMO corporations will seek to extract genes from African indigenous seeds in those crop names and alter the genes so that they can obtain control of these seeds without the consent of the communities who have, over the centuries, cultured and developed these crops as a source of their livelihood; and then claim that the corporations have developed 'new seeds' and 'new crops' out of them. Since they can claim that they are marketing new seeds and crops, they can claim intellectual property rights over these altered natural seeds and prevent anyone from ever planting them without paying them – and can even claim a monopoly right over the seed variety.

What AGRA is doing in going around these farmers rights over their natural seeds is through a programme under the name Agro-Dealer Development Programme, which will provide training, capital and credit to establish small agro-dealers who comprise the primary conduit of the altered seeds, fertilisers, chemicals and the new knowledge to smallholder farmers under the pretext of increasing their productivity and farmer income. In this devious way, the AGRA promoters are a 'grass-based delivery system' where a farmer can 'walk to a shop or kiosk in his rural backyard and readily access high-quality certified seeds' [Odhiambo, A., 2007]. But this is obscuring the fact that AGRA is assisting the GMO agribusiness to establish an entire value chain from 'inputs to markets' that will pave the way for the emergence of a new rural sector: agro-processors and exporters who will contract small farmers to produce crops for them. In short, this will be a short-cut means of transforming the farmers into agents and workers for the GMO agribusiness monopolies such as Monsanto.

Indeed, the AGRA programme has already taken steps towards putting its agro-dealers scheme in place to sell 'improved' seeds, pesticides and fertilisers to poor farmers in East Africa. It has awarded $15 million to a US NGO by the name, Citizen's Network for Foreign Affairs (CNFA), to lay the groundwork and employed people like John Costello, who has a long record in promoting US corporate interests, to head the project. By October 2008 Costello had joined forces with the Croplife Foundation and had announced that they would utilise the AGRA-funded agro-dealers' network, comprising 1500 agro-dealers in Kenya and Malawi, to demonstrate the potential of agro-chemicals. CNFA brought in financial and technical support from other corporate organisations such as Syngenta Crop Protection, Dow AgroScience, Bayer CropScience, DuPont Crop Protection and Monsanto to strengthen the activity of convincing small farmers in Mozambique how to use 'improved' seeds, fertilisers and other inputs 'to expedite their transition from subsistence farming to commercial, quality and maize production marketing' [Mayet, 2010: 12-17].

On its website, CNFA is advertising small farmers and shopkeepers who have succeeded in turning things around by becoming agro-dealers. For instance, in Malawi they tell the story of how one, Dinnah Kapiza, managed to transform a used-clothing business into a chain of agro-dealer shops that form a critical linkage between Malawian smallholder farmers and output markets. In Kenya they advertised the work of Mitunguu Millers Ltd, a small sunflower oil processing business established in 2007 in the Eastern Province of the Maara District. They point out that tackling the product marketing and processing efficiency of a small sunflower oil business in Kenya requires a two-pronged, dual-volunteer approach, which Mitunguu Millers Ltd adopted in Maara where domestic edible oil production exists as a crucial process to smallholder farmer groups and associations. CNFA points out that because of its potential to alleviate rural poverty in Kenya, and as a prominent sub-sector in the country's agriculture, the sunflower seed and oil business has been considered a necessary resource for income by AGRA. They add that the problem is that the vegetable oil market in Kenya is largely dominated by costly imports: primarily palm oil but also soybean, coconut, corn and sunflower oils.

Therefore, the challenge has arisen to create it locally and domestically. They point out that domestic production of the oil must be high-quality, for it risks being phased out in the market. But with current oil extraction and quality below the market standard for cold pressed oils, Mitunguu

Millers Ltd, faces a significant challenge that threatens the company's income. Boosting the efficiency of oil extraction would translate into higher profit margins, a valued product, and penetration of the sunflower seed and oil business into the market. With an increase in production and profits, and further exploration into the environmental impact of sunflower seed production, they add, Mitunguu Millers are now, with the AGRA project, able to capitalise upon further targeted opportunities for the cold pressed sunflower oil, creating a market-driven popularity for sunflower production. The adverts, however, do not refer to the effects that 'improved' seeds and chemicals, which they are promoting through these channels, will have upon the fragile soils of Kenya in the not-too-distant future.

In early April 2011, AGRA announced that in cooperation with the Tanzanian government, smallholder farmers were to receive $7.6 million under the AGRA Breadbasket Project 'to improve productivity and ease access to markets'. It was reported by *The East African* of April 4-10 2011 that the funding will 'benefit farmers who grow staple food crops such as maize, rice and beans' by helping them to buy seeds. The AGRA Market Access Programme director, Anne Mbaabu, announced that about 60 per cent of the fund would go to Ludewa, Mbeya, Mbarali, Kilolo, Songea and Sumbawanga, while the balance would go to other districts, universities, seed companies, breeders and policy hubs. Over $8.2 million of the funding was to go through the National Microfinance Bank and AGRA and the Kilimo Trust were to guarantee small farmers and agro dealers $25 million from Stanbic Bank, which was to assist farmers to purchase inputs and to market their produce under the Warehouse Receipt System.

It was also reported that this breadbasket project will enable farmers in the country to increase their bargaining power; have better storage capacity and access to credit; as well as easing the engagement of contract buyers 'to improve prices to smallholder farmers and assure them a reliable market'. However, AGRA expected the Tanzanian government and donor partners, and the private sector to put an additional $173 million into the project. The AGRA president also announced that AGRA had supported the government to expand a breadbasket strategy and investment plan in Southern Tanzania through the Southern Agricultural Growth Corridor of Tanzania (SAGCOT). But SAGCOT was in partnership with the international global agriculture conglomerates dominating the genetic seed market such as Yara, Diageo, Uniliver, Syngenta, Monsanto, DuPont, USAID, AGRA and the government of Tanzania, which according to the report, had 'already made progress in improving the agricultural contribution to the Tanzanian GDP'.

AGRA, in partnership with the International Fund for Agricultural Development (IFAD), had pooled $160 million 'in affordable loans to agriculture from commercial banks in Tanzania, Kenya, Uganda, Mozambique and Ghana. It was reported that with a budget close to $400 million as of June 2009, AGRA had approved 116 grants valued at $83 million in 14 countries. Grantees of these loans are said to operate across the agricultural value chain, 'laying the basis for the kind of comprehensive, integrated change needed by African smallholder farmers'. It has invested $12 million in 29 projects through the four interlinked programmes of markets, policies, soil and seed since it started its operations in Tanzania in 2007.

The Gates Foundation, working closely with AGRA, is investing heavily in funding research and development of African crop plants. The most important and strategic project is the African Biofortified Sorghum (ABS) project on which the Foundation has spent $16.9 million to promote. The project is spearheaded by a Kenyan scientist, Florence Wambugu, best remembered for her Monsanto-funded sweet potato research, which ended in failure. She has now, under the Gates research project, teamed up with the South African Council for Scientific and Industrial Research (CSIR), DuPont Crop Genetics Research and Pioneer HiBred International to develop a new biofortified sorghum variety, which contains lysine. This will take place at the CSIR in South Africa despite the earlier decision on the part of the GMO authorities to disallow the experiment due to the risk to biodiversity [Ibid.].

Another major project, which has drawn in other funders, is the Biosafety Project. This has been promoted by Monsanto-backed Danford Centre, which will pave the way for the regulatory approval of the GMO crops on the pretext that Danford Centre will provide technical biosafety capacity. Another push is the maize project, which has drawn in a number of Foundations such as the Buffett and Gates Foundations where they are putting into the research a $47 million donation to Water Efficient Maize for Africa (WEMA). This project is being coordinated by the industry-financed African Agricultural Technology Foundation (AATF), which is funded by the British and US governments. It intends to develop a GMO and Non-GMO drought-tolerant maize seed and Monsanto has promised to provide the technology 'free of charge' to WEMA. This 'free' support is, in fact, aimed at obtaining public relations support for its claims that the technology will provide yield insurance, yield enhancement and cost savings on irrigated lands [Ibid.].

The 'confined field trials' of the genetically-modified maize seed were announced to have begun in East Africa in October 2010. The AATF announced that the trials were to begin in Kenya and Uganda in 2010. The research involved the Kenya and Uganda government research institutions, Monsanto, and the Maize and Wheat Improvement Centre (CIMMYT). Together they have developed 12 varieties of WEMA, which are due to be planted in the two countries. It is claimed that the varieties will increase production by between 24 and 35 per cent. It is true that the maize crop that is widely grown in the region is badly affected by drought, but it is doubtful if the technology being promoted will not do more damage to the environment. James Gethi, the Kenyan WEMA-Country Coordinator, in a statement released at the beginning of the trials, said that: 'Everything we have seen in the simulated trials shows that we can safely test transgenic maize varieties in carefully-controlled and confined field trials in Africa [*New Vision*, Kampala, October 18 2010].'

Simulated trials had been conducted in Kenya and Tanzania in 2009. Now the transgenic maize was being planted in one- to two-hectare confined fields, since the Kenyan and Ugandan governments had given regulatory approval, which was given because they were involved in the research. If the maize varieties are approved, the licence is to be claimed by AATF and AATF will sub-license the technology to local seed companies 'royalty-free' for a term or duration to be determined based on future product deployment agreements. Trials are also planned for Mozambique, Tanzania and South Africa in the near future. But more than 30 countries, including the EU, have restricted or banned the production of GM crops because they are not considered proven safe. If this is the case, then it cannot be understood why Britain, which is a member of the EU, should be promoting such a technology for Africans. Clearly, profit for their GMO agribusiness is the consideration.

Rice has also attracted the attention of GMO corporations and multilateral institutions. The African Development Bank (ADB) has launched a $35 million project to support the dissemination of the New Rice for Africa (NERICA), in which AGRA board member, Monty Jones, had played a key role in the research and won a World Bank Food prize in 2004. The project is being coordinated by the African Rice Initiative (ARI), which is hosted by the Africa Rice Centre (WARDA). ARI is mandated to facilitate the dissemination of NERICA across Africa as a contribution towards achieving food security and improving the livelihoods of poor farmers through community-based seed production systems. NERICA is also reported to be performing well in a number of African countries [Ibid.].

But an international NGO called GRAIN has discovered that NERICA threatens the displacement of small farm rice systems with plantation-style rice production managed by agribusiness through the NERICA. The project is based on laboratory research, which ignores small farmers' knowledge and community-based seed systems working with hybrid seeds from the Consultative Group on International Agricultural Research (CGIAR) gene bank. Indeed, it is reported by GRAIN that the small-scale farmers found no favour with NERICA rice varieties and insisted on planting their own seed varieties. This demonstrates that, despite the hype about the 'success' of GMO propagation, there is a growing resistance to hybrid seeds and gene technologies from small farmers. The small farmers still constitute the biggest threat to the agribusiness attempts to monopolise seed production and storage, which may provide the only hope for the world not losing all natural seed production. Thus, the future of a sound agriculture and future prospects for food security lies with the small farmer, who in Africa has the capacity to survive. So it may turn out that Africa did not, after all, 'miss the bus' when they did not join the Green Revolution bus in the mid-1960s. Africa may have survived the onslaught and can now constitute the bastion for resistance against the Green Revolution. But to succeed, they must be reinforced by universities and other institutions of higher learning and research.

What should be obvious, in fact, is that the face of industrial agriculture has changed with the scientific methods that have been used to promote agricultural production along industrial lines backed by reductionist science. The issue is, therefore, for African farmers to take the lessons learnt from the Green Revolution in Asia seriously. These lessons can help African farmers from falling into the same errors that have undermined agriculture at the core. These experiences are [Jhamtani, 2010: 27-30]:

First, that the increase in productivity cannot be maintained in the medium- and long-term. The quick rises in productivity were soon characterised by low rates and falling profitability of enterprises. The experience also demonstrated that the real issue in embarking on systems of increased productivity in agriculture is how to maintain increased productivity while, at the same time, increasing farmers' incomes. If this is not done, the fall in incomes of farmers, which happens when the rapid productivity gives way to low productivity, is debt, suicides and a rapid drift to urban areas.

Second, the Asian experience shows that a high dependence on agro-chemicals may not be a solution to the challenges confronting agriculture in Africa. The Asian experience, based on research by the International Rice Research Institute and the British Department for International Development (DFID), has revealed that insecticides can be eliminated and nitrogen fertilisers (urea) applications reduced without lowering yields.

Third, the use of high-yield seed varieties (HYVs), which are not sustainable in the long-run, creates new problems, which undermine agricultural production at its core. This is because the utilisation of HYV seed tends to undermine plant genetic diversity as monocropping replaces mixtures and rotation of diverse crops, leading to a reduction in resistance to diseases and pests. As the genetic background of HYV crops becomes narrow, their ability to resist diseases and pests declines relative to the ability of diseases and pests to overcome the resistant traits that have been bred into the seed. This requires the recurrent replacement of the seeds, which increases costs and renders HYV crops non-sustainable in the long-run. This situation is being made worse by the drive by Monsanto to seduce and cajole farmers into accepting GMO seeds, which as we have seen, eliminates the natural seeds entirely.

Fourth, intensive double or triple monocropping of rice in Asia has caused degradation of the paddy micro-environment and reductions in rice yield growth in irrigated areas. The planting of paddy rice is increasing on the African continent and therefore African farmers need to learn these lessons from Asia. The problems include increased pest infestation, mining of soil micronutrients, reductions in nutrient-carrying capacity of the soil, building up of soil toxicity, and salinity and waterlogging. This means that areas that have not experienced intensification of the Green Revolution systems, such as Africa, need to undertake a different strategy and technique to achieve food security and this is to be found in the use of organic agriculture [Ibid.].

Fifth and *finally*, whatever agricultural revolution Africa intends to undertake, the Asian experience has shown that Africa needs to take into account the carrying capacity of its rather fragile soils and natural resources and adapt the changes to it. This entails a greater understanding of the physical, biological and ecological consequences of agricultural intensification and greater research attention has to be given to long-term management of the agricultural resource base if the sustainability of agriculture is to be assured. If the issue of the agricultural revolution is to assure people of food security and not profits of agribusiness, then the key question is what kind of technologies are best suited to such an endeavour in the diverse conditions in which African agriculture finds itself.

Quite clearly a 'one-size-fits-all' technological strategy based on the Transfer of Technology model, on which the Green Revolution strategy was based, cannot hold as we have demonstrated. African governments are called upon to develop coherent policies that take the specificity of their diverse conditions seriously in order to promote a sustainable food security on the continent. Africa should learn serious lessons from the Asian experience instead of repeating their mistakes. Indeed, there is evidence that the mainstream scientists are also beginning to listen to the small voices of the farmers in new studies. The *Agriculture at a Crossroads* report, referred to already, has joined the small farmers on this point by emphasising that food security depends on the sovereignty of the small farmer as well as on sound environmental practises. This, the report argues, is critical for the survival of the current and future generations; and that these practises are inexorably tied to ecological agriculture as well as traditional and local knowledge systems, which we are now referring to as *agricology*. It is this regenerative approach to agriculture in its new form of agricology that can ensure that land is not misused through grabbings by agribusiness to turn land into biofuels or agrofuels when the majority of humankind is starving.

iv. From fossil fuels to biofuels – agrofuels

The greed of agribusiness, which was intensified by the global economic meltdown and the crises of the global capitalist economic system, is pushing industrial agriculture to a new self-destructive mode. As we saw above, the drive for super-profits and the resort to financialising had undermined the material production of commodities, which in turn undermined the value of money by the extensive 'production of money' as a profitable activity. Without increasing material commodity production, the financial institutions fell back on certain commodities such as food products to act as 'hedges' against the risk of their valueless instruments such as future contracts and derivatives. This brought speculation into the food production and storage that imitated the financial speculation. Prior to the economic meltdown, the prices of petrol rose sharply to above $150 per barrel around July 2008. All this was due to the general financial speculative fever that had reached its crescendo in mid-2008, which also affected food prices as we have seen above. This development made the search for alternative fuel sources increasingly attractive and the immediate attention was focused on ethanol as the saviour of the world economy.

The new biofuel was derived from plants, which are also food products. There evolved several approaches to meet this demand. One common approach involved the use of sugar cane to produce ethanol from sugar directly. The other approach was to use corn, which was first ground into a fine powder, mixed with water and then heated, after which enzymes and yeast were added to ferment and convert the mixture into a sugar liquid called 'beer' with about a 10 per cent alcohol content. A distillation process then separated the alcohol from the rest of the mixture before the remaining water was disposed of. The remaining substance was essentially pure alcohol. A small amount of gas was added to render the liquid undrinkable. Then the fuel could be used on its own or as a supplement to gasoline to power vehicles as ethanol.

Ethanol was found to have three advantages; at least in theory. The first was that it was, unlike fossil oils, 'renewable'; the second advantage was that it could be produced domestically; and the third was that it burns 'cleaner' than gas or fossil oils. The world's largest producers of ethanol were the US, which made ethanol primarily from corn, and Brazil, which mashed the liquid out of sugar cane. At the time, President George W. Bush and members of Congress expressed support for ethanol use, which was seen as an inducement for the Texas and the Northeast oil monopolies to replace the gasoline additive called MTBE (for methyl tertiary-butyl ether) with ethanol. It was argued that MTBE, which was a chemical used to oxygenate fuel, could contaminate drinking water; while ethanol was judged to present the same danger but was nevertheless said to serve the same purpose in fuel production.

Very soon, other energy multinational agribusinesses, as well as oil and transport companies in the US, European Union and the World Bank, began to promote other forms of agrofuel as a solution to the global energy needs. In this drive, Africa became the centre of attention for the continent was seen as a suitable location of this new 'revolution' in energy production. The reasons given for this preference were many but the most obvious one was that land in Africa was 'cheap' and that it was still 'plentiful' to engage in this form of production, which required a lot of land to grow. Two reports about the phenomenon of land grabs appeared in 2009 and 2010, which drew attention to the magnitude of the drive to grow crops for biofuels.

A World Bank report entitled *Rising Global Interests in Farmland: Can it Yield Sustainable and Equitable Results*, estimated that 21 per cent of these land grab projects were biofuel-driven while the rest were driven by the pressure to secure food security for the countries involved with the land grabs. The report explicitly acknowledged that policies of the developed Northern governments promoting biofuel mandates had played a key role in these developments. It pointed out that: 'Biofuel mandates may have large indirect effects on land use change, particularly (in) converting pasture and forest land, with global land conversion for biofuel feedstocks expected to range between 18 and 44 million hectares by 2030.' This, the report pointed out, would affect the food security of the poor countries.

Another report was issued by the Friends of the Earth movement in August 2010. The report looked at 11 countries in Africa and found that close to five million hectares of land – the size of Denmark – had already been acquired by foreign companies to produce biofuels, mainly for Northern markets. This report entitled *Africa-Up for Grabs: The scale and impact of land grabbing for agrofuels*, was focused on the impact of biofuel demand on land acquisitions in Africa. The report looked at the extent of these land acquisition deals for agrofuels and raised questions about the impacts on local communities and the environment. It found that, although information was limited, there was nevertheless growing evidence that significant amounts of farmland were being acquired for fuel crops, in some cases without the consent of local communities and often without a full assessment of the impact on the local environment.

African leaders were quickly made to accept this political twist and joined in the game of 'attracting investors' who were ready to go into this new venture since it was they who were suggesting the 'investment' in the first place. President Wade of Senegal even went as far as suggesting that agrofuels were a form of 'green energy' in order to make the argument acceptable at home. Wade organised and inaugurated what he called a Green OPEC conference, which was a Pan-African Non-petroleum Products Association (PANPP) composed of 13 countries which did not produce crude oil but which were poised to become exporters of agrofuels products by converting cultivable lands into fuel crop land. Africa was depicted as 'needing help' from this kind of investment because of its low investments from agribusinesses.

Since then, the reasons for the drive for land grabs have intensified as the pressure to produce biofuels has increased. It is now argued that agrofuels are needed as a contribution to climate-change 'mitigation', which means Africa must pay for the greenhouse gas emissions from the Western industries by such 'mitigation' measures. There was also the argument that the world needed 'energy security' by going into this new investment as if this 'security' should be at the expense of African farmers. The shallowest new argument was that agrofuels were important for 'agricultural development', which Africa lacked. For how could one have agricultural development when the effect of introducing biofuels was to undermine the agricultural needs of the farmers, especially their food security?

But very soon, these one-sided arguments and real intentions behind the land grabs for biofuels were confronted with the realities on the ground. The first was the argument that agrofuel crops would be grown mainly on 'marginal lands', which were not under cultivation to produce food. The claim was patently untrue in a number of countries where lands had been acquired under this pretext, as we shall see below. For instance, when land was rented to a Norwegian firm called Biofuel Africa in Ghana, under the claim that it was 'marginal' land, it turned out to be 38 000 hectares of forest and good farming land where the Norwegian company wanted to plant plants for biofuels. It also turned out that some 2600 hectares of this land and forestry were granted to the Norwegians for growing jatropha for biofuel – and this was done before the country's environmental authority had approved the grant. Similar incidents have been reported in several African countries where this has happened on the basis of presidential land 'give-aways' to companies that claimed to be 'investing' in agriculture but specifically to promote agrofuels.

Reports also revealed that in Tanzania, Madagascar and Ghana there had been farmer protests following land grabbing by foreign companies. These companies were accused of providing misleading information to local farmers and of obtaining land from fraudulent community land owners and bypassing environmental protection laws of the countries concerned. The report added that agrofuels were competing with food crops for farmland and agrofuel development companies were competing with farmers for access to that land that would otherwise have gone to food production for domestic consumption. The report especially referred to the case for the jatropha biofuel plant despite the claim that it was being grown on 'marginal', non-agricultural land. It pointed out that by losing the access to their traditional lands, local communities faced growing food insecurity and hunger, and therefore their human right to food was being threatened.

The report also referred to the growing pressure on farmland as a result of these land deals by foreign companies. It observed that the land deals had led to forests being cleared to make way for agrofuel plantations, destroying valuable natural resources and increasing greenhouse gas emissions. In Ethiopia, land inside an elephant sanctuary was said to have been cleared to make way for agrofuels. Farmers had found that the much-vaunted wonder crop, jatropha, rather than bringing a guaranteed income, had in fact taken valuable water resources and was polluting the lands due to the fact that it needed expensive pesticides to grow. In some cases, food crops had been cleared to plant jatropha, leaving farmers with no income and no source of food.

The drive to 'grab' land for biofuels has generated awareness about the adverse impacts of this new form of energy on agricultural production, especially in the food security area. Some of these adverse impacts that have emerged reveal that the people in the affected areas are rarely informed about their unoccupied communal lands being 'given' to foreign companies for this purpose. In many cases, such acquisitions of land have been accompanied by forced resettlements of the population in the lands that were supposed to be 'vacant' or 'marginal'. It has also been found that ethanol production from food crops such as cassava has resulted in the rising of food prices in those countries, thus rendering food too expensive for ordinary consumers to purchase.

In 2008 the International Food Policy Research Institute carried out research which showed that given the continued high oil prices at the time, the rapid growth in agrofuel development was likely to push up global corn prices by 20 per cent by 2010 and 41 per cent by 2020. The prices of oil seeds, including soybeans, rapeseeds and sunflower seeds were also projected to go up by 26 per cent by 2010 and 76 per cent by 2020. Therefore, the introduction of agrofuels has intensified instead of ameliorating the energy and food crises. So where is the solution to these crises? By April 2011, according to the World Bank, food prices went up by over 30 per cent in just four months. This was partly attributed to the rise in fuel prices due to the Libyan civil war. So, it appears that unless the underlying causes of the capitalist economic meltdown are addressed, there can be no end in sight neither to the fuel nor food price increases.

Finally, although it is argued that agrofuels are 'carbon neutral' compared to fossil fuels, in that they do not emit greenhouse gasses, the fact of the matter is that such an argument ignores emissions that are released during production as a result of land-use change, fertilise application and processing in industrial agriculture. All this goes to prove that agrofuels cannot be a solution to Africa's economic development, just as the Green Revolution in Asia and Latin America has shown this not to be the case. Furthermore, the proposed solution to the energy crisis cannot be a long-term solution to the global 'energy security'. In our view, the answer lies in the reassertion of the 'green circular economy', which alone could address the long-term 'soft-energy' path that will transform the economic system. As we shall see in the last sections of this monograph, this path involves a flexible and diverse mix of energy sources and a participation-oriented energy supply structure in both production and consumption – leading to a people-intensive development that can create jobs for everyone [Milani, 2000: 115-6].

[C] The Food Crisis and Land Grabs in Africa

It can now be seen why land grabbing has become such a prominent feature of the post-global capitalist economic meltdown. The search for agrofuels is just one aspect of the attempt to overcome the crisis. The land grabs are being pursued to deal with the other aspects of the economic crisis. To make matters worse, the land grabs are not being pursued by the agribusiness of the developed world alone. They have spread to the oil rich countries in the South as well as the emergent economies. These countries, too, have become active in acquiring large swathes of land in Africa for the food and energy needs of their countries. Indeed, it can be asserted on the basis of the analysis made [2009a, b] that Africa is being caught up in a new wave of recolonisation of its land in order to overcome the accumulation crisis of a late and decadent finance capitalism. It is a form of 'primitive accumulation' for a collapsing system or what David Harvey has called a 'spatial fix'. It is, according to him, a 'spatial temporal fix' intended to relieve finance capitalism from its ills [Harvey, 2003: 87-89].

In those parts of the world, such as Africa, Asia and Latin America, where cassava was the staple food its prices were projected to increase by 33 per cent by 2010 and a startling 135 per cent by 2020. It was further projected that the number of people who will not have enough to eat will rise globally by 16 million for each percentage point increase in the real prices of staple food crops. This means that at least 1.2 billion people worldwide will be suffering from hunger by 2025. If these developments are added to other adverse impacts caused by industrial agriculture, such as the loss of biodiversity due to deforestation, the situation will get even worse. Other adverse impacts will be the non-renewability of agrofuels since biofuels are dependent on finite resources such as land and water. Other effects include the contamination of water sources due to the planting of genetically-modified (GM) crops, which will result in heavy soil quality degradation from the cultivation of jatropha after years of cultivation. If this strategy is to continue, it will lead to a catastrophic ecological collapse of the ecosystem on which all life forms depend, including human life.

As we have already seen, the 2007-2008 global capitalist crises affected not only the big financial institutions and investment banks but also the 'Sovereign Wealth Fund' countries that had invested in the financial assets of these institutions. According to a report from GRAIN, a civil society organisation monitoring land grabs throughout the world, those corporations driving the land grabs are, in large part, 'investors seeking a safe haven for their money amidst crashing financial markets'. They are doing so by buying land cheaply in poor regions of the world, especially Africa, in order to make it economically productive in a short period of time. This would allow them 'to make as much as a 400 per cent return on investment within a few as 10 years' [GRAIN, 2008].

There have been a number of reports over the last two years which have reported on the increasing land grabs in Africa. The first report referred to above and authored by GRAIN appeared in 2008. This report entitled *Land Grab or Development Opportunity? Agricultural Investment and International Land Deals in Africa* was the outcome of collaboration between four international organisations: Food and Agricultural Organisation (FAO), International Fund for Agricultural Development (IFAD), International Institute for Environment and Development (IIED), and the World Bank. The report revealed that over the past 12 months alone, a 'large-scale acquisition of farmland' in Africa, Latin America, Central Asia and Southeast Asia 'had made headlines in a flurry of media reports across the world'.

The report pointed out that lands which had 'a short while ago seemed of little outside interest (were) now sought by international investors to the tune of hundreds of thousands of hectares'. The report added that governments concerned about the stability of food supplies in their countries were promoting acquisitions of farmland in foreign countries 'as an alternative to purchasing food from international markets', which, as we have seen, were highly volatile and speculative. The report warned that these developments would have local and global impacts, especially for food security in the poor countries where the lands were being grabbed.

The second report by the World Bank, already referred to in the earlier section, came out in 2009. The report counted 464 projects that had been involved in the land grabs covering an area of at least 46.6 million hectares of land in Sub-Saharan Africa. By 2011, the total land which was being offered by African governments to foreign 'investors' for biofuels had reached 50 million hectares in total. Another report by the Friends of the Earth, which came out in 2010, also pointed out that, whilst many of these land acquisition deals were for food cultivation, there was a growing interest in growing crops for fuel – agrofuels particularly – to supply the growing EU market. The report also pointed out that these land grabs had been taking place against a backdrop of rising food prices, which led to the food crisis in 2008. There were, as a result, food riots in some developing countries and in Haiti and Madagascar the governments were overthrown as a result of the crisis. The report pointed out that the growing of crops being used for agrofuels was a major factor in the rising prices of foods. It added that as scientists and international institutions were beginning to challenge the climate benefits of this alternative fuel source, local communities – and in some cases national governments – were waking up to the impact of land grabs on the environment and on local livelihoods.

The staggering case is that of land grabbing in Madagascar by a Korean company by the name of Daewoo Logistics, which led to the political crisis in that country. The crisis arose when Daewoo Logistics wanted a large part of Madagascar for its food production. According to the plan this firm had wanted to lease a million hectares of land for a period of 99 years for the production of five million tonnes of corn a year by 2023. The company also planned to use another 120 000 hectares for the production of palm oil. This was reported in a *Guardian* newspaper article, which also revealed that the deal was estimated to cost six billion US dollars over the 25 years. The purchase was acclaimed to be the biggest such deal in the world. This land was estimated to be equal to half the size of Belgium in a country that had itself witnessed food crises domestically. It was also reported that the jobs that were to be generated in this development would have gone to the cheaper South African workers instead of the Madagascan workers.

As the growth in these land deals piled up, reports began to trickle in of attempted 'recolonisation' of African land resources. In Ghana it emerged that some 37 per cent of the entire land mass of Ghana had been 'recolonised' through these land grabs. It was observed that the methods that were being used to acquire these vast amounts of land were reminiscent of methods that 'harked back to the darkest days of colonialism in Ghana and the rest of Africa'. The reports concluded that this meant that foreign companies now controlled 37 per cent of Ghana's land mass for jatropha plantations, pushing small farmers, especially women, off the land. The effect was that valuable food products from crops such as shea nut and dawadawa trees had been cleared to create room for jatropha plantations. In addition, a total of 769 000 hectares had been acquired by foreign companies such as Agroils (Italy), Galten Global Alternative Energy (Israel), Gold Star Farms (Ghana), Jatropha Africa (UK), Biofuel Africa (Norway), ScanFuel (Norway) and Kimminic Corporation (Canada).

An international conference held at the University of Sussex in the United Kingdom on 4-6 April 2011, at which some 100 papers were presented in 32 panels, showed how land, water and other natural resources were being appropriated by national and transnational corporations, as well as by domestic elites and foreign governments through agricultural investments, special economic zones, tourism, conservation programmes, climate change mitigation projects and financial speculation. The papers revealed that not a single case of positive outcomes for local communities had occurred in the form of food security, employment and environmental sustainability. Instead, the conference, which was attended mostly by academics as well as officials from the World Bank, FAO, IFAD, civil society and peasant movements, heard that finance and agribusiness corporations were worried about the impacts of bad press on their reputations.

The conference participants noted that most of the land grabs were speculative: aiming at high returns, which were incompatible with the objectives of food security for the local communities or any legitimate economic activity. The conference participants observed that, therefore, land grabbing only enhances the commodification of agriculture whose sole purpose is the over-remuneration of speculation capital and not the farming communities. Many participants critiqued the so-called Principles on Responsible Agricultural Investment aimed at 'disciplining' land grabbing in an attempt to legitimise something which they considered absolutely unacceptable. The concluding remarks of the conference were very clear that there was overwhelming evidence of the destructive force of land grabbing – for peasant livelihoods and for the environment.

A participant from Indonesia representing a peasant farmers organisation declared that land grabbing was a 'global crime' and exhorted conference participants to build a global moratorium on land grabbing. The International Planning Committee for Food Security and the Land Research Action Network support this view. A UN Special Rapporteur at the conference on the Right to Food observed that trying to make large-scale investments more 'responsible' was not enough. He observed that the real concern behind the development of large-scale investments in agricultural land was that giving land away to investors is resulting in a type of industrial farming that is going to have much less powerful poverty-reducing impacts. It was instead preferable that access to land and water should be improved to the local farming communities themselves. He concluded that accelerating the shift towards large-scale, highly-mechanised forms of agriculture will not solve the problem of hunger: it will make it worse and it was therefore time for Food Sovereignty for the communities.

If the grabbing of land from peasant farmers continues, it will only ensure the realisation of the dream of agribusiness to control all natural resources of the world under a convergence of monopolies called 'agriceutical' super-monopolies. This new monopoly conglomeration comprises a combination of economic interests in agriculture, health, pharmaceuticals, energy and food industries, as we shall see in a section below in connection with the new 'science' called 'synthetic biology'. Therefore, it is important to see how this process of land grabbing from the poor and monopolising of the world's resources is still being organised and envisioned if we are to confront it and ensure that the world's resources truly belong to all humanity and not only to a few individuals that have the power to control the global biomass for a 'new industrial revolution' that is anti-human.

[D] Agriculture and Climatic Change

Climatic change is being taken advantage of by agribusiness and other corporations to intensify industrial agriculture that will increasingly undermine the ecosystem and undermine a balanced global climate. These corporations are pushing genetically-engineered crops as a 'silver-bullet' solution to the ecological and climatic crises to which they have greatly contributed in creating. Thus, instead of admitting the destructive consequences of their profit-seeking activities, they want to dupe the public once more into believing that they have solutions to global warming and climate change crises. In fact, their real intention is to increase corporate power over community-based small-scale agriculture, which alone can revive the soil and the ecosystem if the chemicals that the corporations are pushing down their throats can be stopped. The GMO strategies will result in high costs, reduce the incentive to alternative research and further undermine the rights of small farmers to save and exchange natural seeds. This is in addition to posing serious threats to the environment and human and animal health.

There can be no doubt from the evidence provided so far of the effects of the Green Revolution that this form of industrial agriculture has produced the phenomenon called global climate warming. According to the *Agriculture at a Crossroads* report referred to, about 30 per cent of the global gas emissions that have led to climatic change are attributed to industrial agricultural production mainly emanating from the Green Revolution practises. Climate change in turn has affected all types of agricultural production systems in a way that has led to detrimental development dynamics such as the growing disparities between social and economic groups; the decreasing share of agricultural value added to the global economy; and the degradation of ecosystems throughout the world.

According to the report, results from various assessments of the impacts of climate change on agriculture based on various climate models of emissions scenarios indicate that certain agricultural areas may undergo negative changes. It is estimated that by 2100 parts of the Sahara will likely emerge as the most vulnerable area to climatic change, showing likely agricultural losses of between two and seven per cent of GDP. Western and central Africa will also be vulnerable, with impacts ranging from two to four per cent. Northern and southern Africa, however, are expected to have losses of 0.4 to 1.3 per cent. The forces that shape the climate are also critical to farm productivity. According to the report, already human activity has changed atmospheric characteristics such as temperature, rainfall, levels of carbon dioxide (CO_2) and ground level ozone. These trends are predicted by the scientific community to continue so long as the present practises continue. It is pointed out by weather experts that while food production may benefit from a warmer climate, the increased potential for droughts, floods and heat waves emanating from climatic change are likely to pose challenges for farmers. Additionally, the enduring changes in climate, water supply and soil moisture could make it less feasible to continue crop production in certain regions of the world, especially Africa.

The Intergovernmental Panel on Climate Change, in their 2007 report, have concluded that recent studies indicate that increased frequency of heat stress, droughts and floods negatively affect crop yields and livestock beyond the impacts of mean climate change, creating the possibility for surprises, with impacts that are larger and occurring earlier than predicted using changes in mean variables alone. This is especially the case for subsistence agricultural sectors at low latitudes. Climate variability and change is also likely to modify the risks of fires, pest and pathogen outbreaks, negatively affecting food, fibre and forestry. These studies have also pointed out the different factors and causes of climatic change and agricultural productivity. These are: average temperature increase, change in rainfall amount and patterns, rising atmospheric concentrations of CO_2, pollution levels such as tropospheric ozone, and change in climatic variability and extreme events.

Even these predictions by the IPCC have been questioned by Jim Hansen and his colleagues from the US National Aeronautics and Space Administration (NASA), who have critiqued the climate models used by the IPCC. In their view, these models had even failed to predict the summer polar ice melts that have been happening for years. Jim Hansen regards these models as too conservative and therefore inadequate in predicting the scale of the climate change and its impact on the on-going human activities and natural events [Tayob, T. K., 2010: 6-7].

Other studies have suggested that agricultural production in the US and other industrialised countries is expected to be less vulnerable to climate change than agriculture in developing nations, especially in the tropics, where small farmers may have a limited ability to adapt. In addition, the effects of climate change on US and world agriculture will depend not only on changing climate conditions but also on the agricultural sector's ability to adapt through future changes in technology, food demand and environmental conditions, such as water availability and soil quality. Management practises, the opportunity to switch management and crop selection from season to season, and technology can help the agricultural sector cope with and adapt to climatic variability and change.

But these predictions ignore the global implications of climatic change and the demands that the developed world imposes on less developed parts of the world and the impact of those impositions on the economies of the developed world. In fact, recent developments in food and energy prices, as well as the global financial crisis, as we have seen, have demonstrated the interdependence of the world, hence the need to think global and act locally in a Glocal new world. The impacts of the financial crisis have created demands on the other continents, such as Africa being required to provide land to meet the food and energy needs of the other parts of the world; and these demands impose new constraints on African small farmers and their capacity to survive climatic change.

Recent weather conditions have demonstrated the global nature of the climatic warming to be a reality for everyone, including the developed world. Formerly, it was argued that only a certain category of countries such as the Island countries and least developed countries were the most vulnerable to climatic change. But because of the high temperature conditions throughout the world, countries that were not regarded as 'vulnerable', such as Pakistan, Nicaragua and other Central American countries, which have been affected by hurricanes due to high temperatures, have claimed that they, too, are extremely vulnerable and have questioned the criteria for categorising countries in this way.

Indeed, the 2010 heavy rains and flooding that visited on China and Pakistan have gone far in proving that climate change is behind these occurrences. These rains have, in the case of Pakistan alone, affected some 20 million people. Nine hundred thousand homes were damaged or destroyed and 4.6 million people were rendered homeless just in two provinces of that country. As a result, 6.5 million people were without water, food and medicines. In Khyber Pakhtunkhwa province alone, 70 per cent of the bridges and road works were destroyed. Not only was human life badly affected, the rains also destroyed agricultural crops and animals, including cattle. The scale of the natural calamity in Pakistan was said to be worse than that occasioned by the Asian tsunami and the Haitian earthquake tragedies combined.

Despite those local factors that were attributed to the floods, such as the chopping of forests, mismanagement of rivers and land, etc., the fact of the globality of these events should not be lost on all of us. The Geneva-based World Meteorological Organisation confirmed that these floods in Pakistan and China were connected to global climate change, which confirmed other observations and predictions. Moreover, the 2010 high temperatures that have prevailed in this year's summer in countries such as the US, Europe and Central Asia also go to prove the depth of the change in the world weather conditions. This also explains the prevalence of the stronger-than-usual monsoon season in South Asia.

The Intergovernmental Panel on Climatic Change (IPCC) has observed that globally around 50-80 per cent of organic carbon that was once in the topsoil has been lost to the atmosphere over the last 150 years or so, due to our failure to take care of the earth as a living organism. By inference, the IPCC observes, degraded soils have the potential to store up to five times more organic carbon in their surface layers than they currently hold, 'provided we change the way we manage the land'. There should, therefore, be no complicacies about which part of the world we live in. Indeed, the IPCC continues to observe that when appropriate changes to land management are put in place, agricultural soils will have the capacities to sequester and store large volumes of carbon, thus improving microbial content, biological activity, fertility, structure, stability, resistance to erosion and lead ultimately to biodiversity, productivity and profitability. Therefore, according to Jones [2006], increasing soil carbon can significantly reduce the impact of dry land salinity, reduce sedimentation rates in rivers and streams, improve water quality, improve air quality and decrease the impact of the Greenhouse Effect, global warming and climate change [Ibid.].

To achieve these levels of reversal of the impacts of climatic change requires global action. For instance, a complete ban on chemicals and reduced livestock units per acre, as is practised under the organic farming systems, has the potential to reduce the concentration of easily available mineral nitrogen in soils and thus N_2O emissions [Dharmitra, 2009]. The issue of land losses and arable croplands and permanent pastures caused by mineralisation, erosion and overgrazing can also be combated through similar means.

The *Agriculture at a Crossroads* report has explored the demands that are likely to be made on agricultural systems in view of these developments in the form of crops, livestock and pastoralism, fisheries, forestry and agro-forestry, biomass, commodities and ecosystem services in the future. The report asks what agricultural goods and services society will need under different plausible future scenarios in order to achieve the goals related to hunger, nutrition, human health, poverty, equity, livelihoods, and environmental and social sustainability. It also asks whether and how access to these goods and services will be hindered. The result of this analysis is an evidence-based guide for policy and decision-making for governments and economic actors. This analytical requirement points to the need to address these challenges through a different designed agriculture that adapts to climatic change impacts as well as to the reduction of greenhouse emissions. In this connection, there have been two buzz-words in regard to interventions to deal with climatic change. These are 'mitigation' and 'adaptation', which are recommended to be applied in a generalised manner.

For some time, most debates about how to deal with climatic change have focused on 'mitigation' and ways in which greenhouse emissions could be reduced. But the critical situation is on the vulnerability of marginalised communities who had to adjust their agriculture and water management practises to the adverse conditions created by industrial agriculture. In short, they have been asked to cope with these adverse changes in the global climate systems although they are least able to do so. This has raised the issue of 'adaptation' of these communities to global climatic change and what that entails on both a local level and in terms of institutional and policy changes. In effect, this demonstrates the peculiar situation in which the marginalised communities are being asked to 'adapt' to problems that had been created by the big corporations from the developed world and the lack of an effective framework to address these problems at a global level.

In effect, the adaptation approach came to mean that the marginalised local communities should develop local autonomous adaptations, while support was also requested to enable them to 'bridge the information and funding gaps that impeded their present capabilities to adapt to climatic change and to share successful strategies with others in similar situations' [Both Ends, 2007: 3]. While this meant that these new mechanisms were needed to channel funding effectively to people that needed it most, it also became clear that sustainable adaptation had to focus on building the adaptive capacities of communities for sharing and strengthening existing knowledge about climatic variability, its impacts, and local adaptation strategies. In addition, knowledge-sharing links need to be established between local adaptation needs and actions and policy and investment processes on national and international levels [Ibid: 9].

At local levels, this could be met through sustainable and integrated agricultural systems and by putting in place supportive policies and programmes for such systems to be implemented. For instance, although agriculture itself contributes to climatic change with around 10-12 per cent of the global anthropogenic greenhouse gas emissions annually, this is mainly from methane produced from livestock raising, biomass burning and wet cultivation practises, as well as from nitrous oxides from the use of synthetic fertilisers. If the emissions from the production of synthetic fertilisers and the total food chains from the farm to the consumer are considered, the greenhouse gas emissions from all sectors related to agriculture add up to the 25-30 per cent mark as pointed out by the *Crossroads* report. The challenge is, therefore, to address these problems at source. It is only a combination of local adaptations and global responses aimed at overall adaptations that can deal with the problem significantly – hence the need for 'glocal' responses. This is the challenge facing humanity caused by global climatic change.

[E] The Destruction of the Small Farmers

The consequence of capitalist scientific agriculture has been an attempt to destroy the small farmer and livestock keeper who has historically produced and reproduced natural seeds and domestic animal husbandry. They have, through centuries, provided food for the world's population as well as for their families. This historical fact was acknowledged by the United Nations Secretary General, Ban Ki-moon, at the 2010 Food and Agricultural Organisation (FAO Summit). For instance, almost 80 million smallholder farmers in Africa supply about 80 per cent of its food supplies. In order for Africa to establish a prosperous and sustainable economic future, the voices of its small farmers must be amplified.

As we noted above, right from ancient times, farming was based on the small farmers reproducing themselves on the circle of seed planting and replanting. This power of the small farmer has been increasingly undermined by modern industrial agriculture. The Green Revolution in India was the first to witness the dispossession of the peasant farmers in the global South due to indebtedness caused by the agricultural changes. This meant the small farmers had to leave the countryside and join the jobless and landless living in the shantytowns such as Bombay and New Delhi. This has been the general trend in almost all the countries of the world.

According to the Rockefeller Executive Director of the Agriculture Development Council, the objective of introducing new systems of production among the small farmers was to teach the peasants to 'want more for themselves' and to abandon 'collective habits' in order to get on 'with the business of farming'. Of course to do this they had to raise credit in order to buy the inputs for the new form of farming for the market, which impoverished them even more. The Green Revolution provided the conditions where the small farmer had to borrow at extortionate rates from local money lenders to whom they were permanently indebted in order to keep going deeper into debt. These money lenders were the main dispossessors of the peasant lands in addition to the economic activities arising from the revolution itself. According to Engdahl [2007: 128]:

> 'One major effect of the "Green Revolution" was to depopulate the countryside of peasants who were forced to flee to shantytown slums around the cities in desperate search for work. That was no accident: it was part of the plan to

create cheap labour pools for the ... US multinational manufacturers [p. 128].'

We have seen above that the groups who gained most from the Green Revolution in India were the rich farmers and pockets of agribusiness tied to large export monopolies such as Cargill. The regions where the vast majority of poor peasants worked remained poor and the gap between the rich feudal landowners and the poor peasants remained high. This trend was also observed in Mexico where the Green Revolution was first tried out. Here, too, the high-yield variety seeds were planted in the rich areas reinforcing the old semi-feudal *Latifundista* divisions between wealthy landowners and the poor peasant farmers who were similarly dispossessed as a result of the new farming methods. Although working under different conditions, the US small farmers did not fare any better but worse. The strategy of the Rockefeller Foundation on behalf of the other US agricultural monopolies was first to extend their markets overseas. It was the first step in a well-planned process of dispossession of the small farmers. This trend was later dubbed 'market-oriented agriculture', meaning a move towards an agribusiness monopoly. Having captured these markets outside the US, these monopolies turned inwards in the US itself to reorganise US agriculture towards a concentration of farmlands under their control.

The trigger to this process was the increasing decline in industrial manufacturing production. The application of science and technology to the manufacturing industry had led industry to the general loss of profitability of firms due to the costs of labour increasing beyond profitability. This was expressed in terms of US investment in industry going abroad to lands where labour was cheaper. Hence, US jobs in the traditional industries such as iron and steel as well as textiles disappearing abroad in Southeast Asia. This was called 'deindustrialisation' and 'relocation' into cheap labour countries. According to Engdahl:

'American industry was rusting as its factories, most of which were built before and during the war. [These industries] had become obsolescent compared with the modern new post-war industry in Western Europe and Japan. Corporate America faced severe recession and its banks were hard-pressed to find profitable areas of lending [Ibid: 38].'

The result was that US corporations were forced to reorganise internally to regain profitability in other areas and agriculture was the only recourse left to big capitalist investors. Already by the 1960s, food or agribusiness was becoming an area of US global domination and extensive investment in research had occurred, part of which went into the Green Revolution as we have seen. Petroleum was another area, which had already been monopolised by the form of the 'Five Sisters' oil cartel. Even here, Rockefeller had led the way through their vertically-integrated Standard Oil Trust oil cartel, joined by four other 'Sisters' in fixing prices of petrol according to their profitability calculations. In the area of agriculture five food 'Sisters' had also emerged; these were accused by the US Federal Trade Commission of trying to 'monopolise the entire nation's food supply' by the 1920s. They had acquired a near monopoly in meat packing and thereby controlling access to public stock yards of cattle. The five agribusiness monopolies, following the five oil sisters also adopted a 'vertical integration' strategy for the meat industry in which they 'integrated forward' into the marketing of the beef, and backward into the supply of raw material feeds for beef cattle and hogs [Ibid: 135].

With the crisis of profitability in industry there was no going back; and the Rockefeller and Ford Foundations promoted the 'vertical integration' strategy as a general business strategy by funding a research project at Harvard University called the Harvard Economic Research Project on the Structure of the American Economy, which was headed by three economists, Leontief, Goldberg and Davis. The project produced a theory of economic management that re-introduced vertical integration in US food production. This led to a pressure to 'deregulate' and 'privatise' the economy under the regimes of Margaret Thatcher and Ronald Reagan in the UK and the US under the ideological slogan of 'freedom' and 'free markets'.

This was indeed a paradigm shift because the process under the new theory led to concentration and transformation of American agriculture as a whole to make it more 'efficient'. The concentration led to the independent family farmers being driven off their land to make way for the now efficient, giant corporate industrial agribusinesses. Agriculture became more and more industrialised as the manufacturing industry became increasingly unprofitable. Under concentration, farming became a big time operation under Factory Farms or 'corporate agriculture', which increasingly took over the land from family farmers who were now forced to work for the new monopolies as 'contract farmers'. The previous regulatory framework, which had been put in place to protect family farms on food safety and food security was loosened and even removed during the 1980s – the period of the neoliberal economy. This opened the doors to financial investment banks such as Goldman Sachs to enter the food market to speculate on food products resulting in the 2008 food crisis to hedge against the risks of the new financial instruments such as derivatives.

Agribusiness and industrial agriculture began to transform the face of traditional American farming in ways so drastic as to be incomprehensible to ordinary consumers. Many consumers still thought the farm products they were buying were from family farms. What was in fact happening was that the wholesale merger and consolidation of American food production out of the hands of the family farmers and into giant corporate global concentrations was taking place. The family farmers gradually became contract employees responsible only for feeding and maintaining concentrations of thousands of animals in giant pens. They neither owned the farms nor the animals. They had become something comparable to feudal serfs and indentured labourers through huge debts [Ibid: 137].

In this way, hundreds of small family farmers were forced out of farming with the spread of agribusiness and its large operations. This was because traditional family farming was labour intensive by its nature, while factory farming was capital intensive. In time, the increase in factory farms led to a decrease in the price which independent farmers got for their animals. This forced thousands of farmers still struggling to survive on the farm out of business. The number of farmers in the US dropped by some 300 000 between 1979 and 1998, while the number of hog farms decreased from 600 000 to 157 000, while the numbers of hogs sold increased. The new giant Factory Farms were estimated to be killing three traditional farm jobs for every new, often lowly-paid, job they created. As the share values of family farms dropped, those of corporate agribusiness increased. The farm subsidies that had originally gone to the family farms were now claimed by the new agribusiness monopolies [Ibid: 143].

By the 1990s, the ability for corporations to merge and vertically integrate had created a structure never seen before. By the end of the 1990s four large corporations controlled 64 per cent of all pig packing in the US and another four controlled 84 per cent of beef packing. Another three controlled 71 per cent of soybean crushing and flour milling. Two giant agribusiness corporations – Monsanto and Pioneer HiBred of DuPont – controlled 60 per cent of the US corn and soybean seed market, which consisted entirely of patented genetically-modified seeds. The new agribusiness sector had become the second most profitable 'industry' in the US, next to pharmaceuticals, with annual domestic sales of well over $400 billion. The next phase in the vertical integration game was the merger between agribusiness and the pharmaceutical industries into a giant monopoly. The interests of the State and agribusiness had also become closer so that on the eve of the Iraq war in 2003, the Pentagon's National Defence University had the courage to declare that: 'Agribusiness is to the United States what oil was to the Middle East' [Ibid.].

The merger of agribusiness with the State meant that all the barriers that had been created in the defence of the small family farm were now removed and new ones erected to protect agribusiness. As we have already noted, all subsidies formerly paid to small family farms were not paid to them. Even economic theory and policy changed in their favour. The farm policy that had been put in place under the Agricultural Adjustment Act of 1938 had granted authority to the Secretary of Agriculture to balance supply and demand for agricultural products by idling land during abundant supply. This was intended to implement commodity storage programmes as well as establishing market quotas for some crops in order to encourage exports of commodities abroad including food relief programmes and sales of farm commodities for soft currencies. This authority was suspended in 1996 and finally repealed in the 2002 Farm Bill to the detriment of family farms.

The expectation behind the new farm policy was that the 'market forces' would appropriately allocate resource use in agriculture. In fact what happened was that 'the market forces' instead allocated the new agribusiness huge opportunities to purchase large tracts of land at cheap prices. Therefore, what agribusiness appeared to 'lose' in the cheap prices of agricultural produce, they gained by acquiring more land from the farm families, only later to hike the food prices once they had gained the monopoly over the market throughout the world. A report done by the University of Iowa concluded that:

'When no land is idled, production increases, crop prices fall, and land values come under pressure until there is less profitability for crop production on the least productive land. The market squeezes out the thinner soils and steeper slopes, the higher per-unit cost of production areas. This land then transitions … to another crop or to grazing land [Quoted Ibid: 145].'

What is suggested here by this report is that complete dispossession of the small family farmers of their land had led to the cheapening of its price (which reflected itself as a cheapening of their products) and that such a cheapening of the value of the land had led to the further dispossession of the farmers as the value of their land came under pressure. This continued until there was less profitability for crop production on the least productive land owned by the family farmers. In that case, the family farmer could no longer farm at those low 'less than profitable prices' and therefore he was forced to sell the land to agribusiness. This is then what enabled the 'transitioning' to take place to another crop or for the land being relocated to grazing and to a complete monopoly over land by agribusiness. Having achieved the monopoly over land, they were able to determine new levels of costs of production and monopoly prices for the products through 'vertical integration'.

As Karl Marx had predicted, land now had *a new monopoly price* so that the general price of production of agricultural commodities was now determined by the *cost of production on the worst piece of land plus an average rate of profit*. This was a radical departure from price determination in the manufacturing industry under competitive conditions, where the price of a commodity was based on a social average of the cost of production that prevailed in the entire sector of the economy. This departure in price determination was justified on the basis that the 'naturally based' differentials in productivity on land could not easily be eliminated by technological change in the same way it could in industry. *It was also argued that an expansion in agricultural production entailed drawing more inferior lands into cultivation as well as intensifying production on superior soils when it was more profitable to do so* [Harvey, 2006: 342 – emphasis added]. This was the only means through which monopoly capitalism in general could survive in the world by dispossessing the poor famers in order to replace a failing manufacturing industry sector with monopoly agribusiness, which, as we shall see below, was being forced to structure even more towards the control of biomass as a basis for the 'new industrial revolution'.

In short, although in the case of the United States, unlike Europe, the dispossession of the aboriginal peasant farmers had taken place, the vast stretches of land that had been dispossessed from them had been taken over by small capitalist family members who, under competitive conditions, had managed to make a living without big farm monopolies. Because of that the emerging agribusiness monopolies could not establish a new system of monopoly pricing over the products of the land until they had fully attained the control over the total agricultural land from the family farmers. This was under the neo-liberal economic policies pursued by the Thatcher and Reagan regimes in the UK and the US between the mid-1980s to end of the 1990s.

The Harvard group of the vertical integration project had predicted that the addition of entire new sectors created by the latest developments in genetic engineering, including GMO and the creation of pharmaceutical drugs from genetically-engineered plants, which they now called a 'convergence' into the 'agri-ceutical system', was likely to create a new, huge market. Vertical integration would have eliminated competition in those sectors. Goldberg, one of these agronomists, argued that:

'The addition of life science (biotechnology) participants in the new agri-ceutical system will increase total value added in 2028 to over $15 trillion and the (small family) farmers' share will shrink even further to 7 per cent [Ibid: 146].

In that case, Goldberg predicted that 'the genetic revolution' would lead to an 'industrial convergence' of food, health, medicine, fibre, and energy business into a single gigantic monopoly business activity for the next century. How the new genetic revolution was to evolve depended on how the promoters of this revolution, especially the Rockefeller Foundation, were to face the task. The issue was not how to move from the Green Revolution to the Gene Revolution, it was rather how to transform the way people of the world were to feed or not to feed themselves with their loss of control over land and seeds. There was no going back for both the agribusiness and the small farmers: either agribusiness would succeed or the small farmer would strive to survive through struggle. A new super-class was emerging globally that was to ensure that the agenda of agribusiness was achieved, just as small farmers were strategising how to resist the new enslavement.

[F] FROM THE OLD INDUSTRY TO THE 'NEW' BIO-INDUSTRIAL ECONOMY

The crisis of productivity in industrial agriculture and the degradation of the ecosystem as well as the need to adapt to the impacts of climatic change in agriculture, has prompted a number of agribusiness corporations to embark on new 'scientific' strategies aimed at putting in place a 'new industrial revolution'. This has led to research in 'new crop varieties', which in effect will privatise the common property in the seeds of society without their consent. The new strategy will also lead to the privatisation of all biomass of the world by a few super monopolies. As we shall see below, this is being done through the introduction of a 'new science' called 'synthetic biology' and 'nanotechnology', which will supplement molecular biology and genetic engineering in appropriating biomass.

i. A 'new science' for a 'new system'

These developments are being done under the pretext that these corporations, whose only concern is the search for super-profits for themselves, are developing 'climatic-ready' and 'drought-tolerant' crops in the form of genetically-engineered crop varieties to reduce the effects of global warming and ecological degradation. Many of these corporations are trying to privatise and patent commercially viable traits of the existing natural seeds and plants that can adapt to the impacts of climatic change and turn them into their monopoly property rights to be sold to the real small farmer owners of the natural varieties from which the 'new varieties' are being 'engineered'. This is intended to ensure that as the climatic change continues to bite these corporations will be assured of markets for the 'new variety' seeds and their products, which will be protected by patents under the World Trade Organisation rules.

In fact, the on-going economic, environmental, social and political crisis around the world has created a fundamental problem, for the corporate world is becoming unmanageable for the States, although the corporations are trying to resolve some of the economic and environmental aspects; but their strategies will make the crises worse than before. The strategies they are working on are increasingly converging around the control and exploitation of *biomass* as a possible solution to the economic crisis. But what is biomass? Strictly speaking, biomass is a measurement of weight that is used in the science of ecology. It refers to the total mass of all living things or organic matter found in a particular location. Fish, trees, animals, bacteria – and even humans – are all capable of being referred to as biomass. However, more recently the term is being used as shorthand for non-fossilised biological materials, particularly grass and plant material that can be used as a *feedstock* for fuel or for industrial chemical production.

According to the United States National Renewable Energy Laboratory, biomass includes 'organic matter available on a renewable basis', such as forest and mill residues, agricultural crops and wastes, wood and wood wastes, vegetable oils, animal wastes, livestock operation residues, aquatic plants, fast-growing trees and plants, and municipal and industrial waste. On closer examination, however, the latter includes motor tyres, sewage sludge, plastics, treated lumber, painted construction materials and demolition debris. Even industrial animal manures, offal from slaughterhouse operations, incinerated cows and landfill gases all seem to fit the description of biomass.

Historically, biomass has been a source of fuel and material production for millennia. However, the new use of the term 'biomass' marks a specific industrial shift in humanity's relationship to nature. Unlike the term 'plants', which opens to a diverse taxonomic world of various species and multiple varieties, the term 'biomass' treats all organic matter as though it is the same undifferentiated 'plant-stuff'. This is rather a crude modern understanding of natural life, which is consistent with the modern notion of how to deal with nature as 'a resource for exploitation'. Recast as biomass, in the new usage, plants are simply reduced to their common denominator so that, for example, grasslands and forests are redefined commercially as simply sources of cellulose and carbon.

In this way, biomass operates as a narrower and profoundly reductionist and anti-ecological concept that treats plant matter as though it were a homogenous bulk commodity, which is available for exploitation as a 'commodity' or a 'resource'. The use of the new concept 'biomass' to describe natural substances and living organic matter is therefore going to push nature to a new corner that can be exploited at will, but with dire consequences to the future of humanity.

The crisis that the world is facing, since the 2007-2008 economic 'meltdown' has been especially felt in the area of rising energy and food prices. But this was merely a surface reflection of the deeper cancerous nature of late global capitalism. Suddenly new buzzwords such as 'bio-fuels', 'agro-fuels' and 'land grabs' began to be heard with ever-increasing regularity throughout the world as the corporations tried to seek ways of overcoming the crisis. The embarking on agro-fuels signalled for a call for a 'green economy' by mainstream United Nations institutions such as the United Nations Environmental Programmes (UNEP). The call for such an economy was an attempt to underscore the level of the crisis as lying in the exhaustion of natural resources or their loss of capitalist 'profitability' to the capitalist industry. It envisaged a fundamental transition in industrial strategy towards something new in its exploits. That new trajectory, variously called the 'new bio-economy' or the 'bio-based economy', was gathering momentum as well as political clout. Billions of dollars in public subsidies and private investment were pumped into the 'green' energy sector as well as land acquisitions to ameliorate the crisis by producing biofuels.

This fundamental 'swift' in strategy was taking place in the context in which biotechnology had taken the centre stage in industrial agriculture with genetic engineering at its core. Indeed genetic engineering had already, with molecular biology, created a convergence of a number of economic sectors into one gigantic monopoly called the agri-ceutal industry. This convergence had combined genetically-engineered agriculture, plant-based pharmaceuticals, energy, health industry and finance into a co-related activity. This combination had become a major industry in its own right after the decline of the manufacturing sector. A new reductionist science in the name of 'synthetic biology', had, within this short time, emerged to be the basis of the new combined industry taking stock of the agri-ceutical economy. The emergence of synthetic biology in many ways was an acknowledgement that molecular biology, which had sponsored genetic engineering, had exhausted itself just as eugenics had earlier done leading to molecular biology.

The problem was that the crisis of industrial capitalism had also affected the way science was being used to address its crises. The new science of synthetic biology was contrived on little empirical evidence; but on analogies and assumptions drawn from the failed molecular biology. Although it was being referred to as a 'maturing scientific discipline' that combined science and engineering, its scientific basis and foundations were indeed shallow. Back in 2000 the science was referred to by the majority of scientists in Canada as 'a decision procedure for facilitating the passage of new products – GE and non-GE – through the regulatory process'. An organisation called the ECT Group has recently written a monograph entitled *The New Biomassters: Synthetic Biology and the Next Assault on Biodiversity and Livelihoods* [2010], in which the group describes how synthetic biology has ballooned itself into becoming an important 'new science' from the 'fringe' science it had been, into a 'hybrid science' combining engineering and computer programming. What marked its rise to prominence in a short five-year period was the intense industrial and investment interest it had generated and excited in the failing industrial and agribusinesses. This demonstrated that global finance capital was enveloped in an urgent quest to find new avenues for quick investment and recovery.

That is why synthetic biology is a contrived technique claiming to be a science: because of the big economic interests that were behind its creation. Thus, although it claims to be a new science it was in fact contrived from the old reductionist sciences and techniques to meet a particular purpose. It is not a science in the service of humanity as a whole but a technique that is intended to serve the interests of the old agribusiness and industrial monopolies out of their crises. It is intended to 'combine' old reductionist systems, which had destroyed nature and material production into a new anti-nature exploitative system. It is therefore not based on a *holistic* understanding of nature but on assumptions and analogies drawn from a reductionist, failed, molecular biology. Its central thesis is that a DNA that is drawn from a sugar-based molecule consisting of four types of chemical compounds can be organised in a unique sequence to form a code that can instruct living organisms how to grow, function, and behave. In short, synthetic biology tries to imitate nature and rewrite its DNA code to get it to act in the way the new science wants it to. According to the ECT Group:

> 'By rewriting that code, synthetic biologists claim they are able to programme life forms much like programming a computer. These assumptions are based on a model of genetic systems that is over 50 years old, known as the "central dogma" of genetics. However, the accuracy of that dogma is becoming less and less certain [Ibid: 37].'

In other words, synthetic biology is not based on empirical scientific evidence but on reductions, which themselves are also not based on scientific empirical evidence. The reductions are in fact drawn from other assumptions based on molecular biology, which was used to 'prove' that a product produced by means of genetic modification can be 'substantially equivalent' to a natural product such as a natural seed. This is the *central dogma* that backs the 'scientificity' of synthetic biology. This dogma is not drawn from a scientific discovery based on evidence. On the contrary, it was a *political dictate* in the form of a political *Executive Order or Directive* from President George H. W. Bush – the father of George W. Bush – in an attempt to assist the agribusiness in their drive to introduce and market GMO seeds around the world. Under the threat of those who objected to the marketing of GMO products as being dangerous to human and animal health and unable to demonstrate that the GMO seeds and products were safe for human consumption, the agribusiness approached President Bush to assist them.

Therefore, if science could not help agribusiness to legitimise their products, the highest office in the land – and globally – would do what science had failed to accomplish. According to President Bush, he was approached to 'solve the problem' by declaring that GMO plants and foods, which were being marketed by the GMO agribusiness were *as good as* the natural products. This is what came to constitute the central dogma, which states that GMO products were 'substantially equivalent' to ordinary (natural) plants and foods if they were 'of the same variety' as the natural ones [Engdahl, op. cit. 7]. Later the dogma was adopted and generalised by the OECD to mean that a novel genetically-modified food should be considered to be the same as the natural food if it demonstrated that it had the same variety and composition as the conventional food. This is the dogma and not the science that synthetic biology was trying to sell to the world.

The concept was coined in the case of the US for marketing purposes and for the OECD for regulatory purposes. In the case of the US, the designation of the product as 'substantially equivalent' was intended to authorise US regulatory bodies not to require the agribusiness to carry out scientific tests in the form of biochemical or toxic tests on the GMO products since it had been declared by the top executive to be 'substantially equivalent'. The acceptance by the OECD helped to generalise the dogma and to create what is now called 'international consensus' on the 'principles' regarding evaluation of the food safety of genetically-modified plants. It is based on the idea that existing foods can serve as a basis for comparing the properties of genetically-modified foods with the appropriate counterpart, with the implication that the modified product can be regarded as having the same chemical composition and nutritional value. This is, in fact, an attempt to reverse the natural process of seeds and plants so that a reductionist technique can claim to have the scientific right to dictate how nature should 'grow, function, and behave'.

However, the downside is that even then the application of this dogma is not accepted by the scientific community as a whole as a safety assessment technique *per se*. It is regarded as merely helping to identify similarities and differences between the existing natural foods and the new genetically-modified products, which should then be subjected to further toxicological investigation. 'Substantial equivalence' is regarded by these scientists as a starting point in the safety evaluation, rather than an endpoint of the assessment. The dogma cannot, therefore, be used as a basis for developing a new scientific theory. It is merely a bureaucratic dictum that has been elevated into central dogma in molecular biology, which synthetic biology has adopted. It cannot even be used as a hypothesis for testing a scientific inquiry. Indeed, a Global Network of Physicians and Scientists for Responsible Application of Science and Technology based in Canada has called dogma 'scientifically invalid'. So there is a hope for humanity so long as a holistic science that takes into account the needs of nature and humanity converges for the future.

The Royal Society of Canada concluded, as we noted above, that the dogma 'of substantial equivalence does not function as a scientific basis for the application of a safety standard but rather as a decision procedure for facilitating the passage of new products, GE and non-GE, through the regulatory process'. This Global Network of scientists observed that although the dogma was questioned way back in 2000, 'yet this unscientific and dangerous approval procedure is still being used, although 10 years have passed'. The scientists further observed that this indicated that the *governmental authorities dealing with approval (procedures) are ignoring science in a seriously irresponsible way*. They added that these governmental institutions included the Food and Drug Administration (FDA) in the US and the European Food Safety Agency (EFSA) of the European Union, among others. This leaves us with no option but to conclude that the synthetic biology on which these unscientific procedures are based is a contrived technique that has no scientific basis.

Contrived as it is, synthetic biology is being used to create a new 'industrial revolution'. Its main objective is to *repurpose* simple cells such as yeast or bacteria 'to behave like factories'. It is being used to design and construct 'new biological parts', 'devices', and 'systems', as well as the re-designing of existing, natural biological systems for 'useful purposes'. This redesigning includes photosynthetic systems to produce energy. It is therefore used as an engineering device of biology, which is used for the purposes of *synthesising* complex biologically-based or biologically-inspired systems, which display functions that do not exist in nature. This engineering device can be applied at all levels of the hierarchy of biological structures – from individual molecules to whole cells, tissues, and organisms to produce the desired results. In essence, synthetic biology is a device that enables the designing of biological systems in a 'rational' and systematic way.

There are other reasons that have led to the rapid emergence of synthetic biology as a new knowledge paradigm that is cutting across other sciences and procedures. Looked at as a science, it is undermining many previous boundaries that existed between the different sciences and between the natural sciences and social sciences. This has resulted in a narrower specialisation within the biological sciences, which has at the same time captured the imaginations of some scientists from other scientific disciplines in related fields which apply methods used in non-biological fields like mechanical engineering, electrical engineering, and computer science to configure biological systems to achieve important practical purposes. This development has been referred to as 'genetic engineering on steroids', which now dominates discussions of synthetic biology and, according to the ECT group, constitutes a new *sensu stricto* definition of the term.

This development is perhaps the reason why synthetic biology is also being referred to as a set of 'extreme genetic engineering' devices and techniques. The techniques involve constructing novel genetic systems by *applying* engineering principles and synthetic DNA to achieve certain results. In that way synthetic biology techniques differ from transgenic techniques that 'cut' and 'paste' naturally-occurring DNA sequences from one organism into another in order to change an organism's behaviour, such as putting bacterial genes into corn or human genes into rice. Synthetic organisms so created are machine-made life forms or living organisms such as yeast or bacteria to which strands of DNA are added. These organisms are constructed by a machine called a DNA Synthesiser, which uses the techniques of synthetic biology. Synthetic biologists can thus build their DNA from scratch using this machine, which can 'print' the DNA 'to order'. In this way, the biologists are able to radically alter the information encoded in the DNA, creating entirely new genetic instructions as well as jump-starting a series of complex chemical reactions inside the cell, known as metabolic pathways. In effect, the new, synthetic DNA strands can 'hijack' the cell's machinery to produce substances which are not produced naturally [ECT Group, 2010: 36].

But a more broad-based holistic scientific research in genetic science, especially in the field of 'development systems theory' and 'epigenetics', has questioned the prominence being attached to synthetic biology's DNA code. These scientists have pointed out that all manner of complex elements both within and outside a living cell influences the way a living organism develops and this cannot be determined *a priori* by focusing solely on the DNA code. They also point out that even environmental factors, such as stress and the weather conditions, can influence their development. Accepting this critique, some synthetic biologists have admitted that their carefully-designed DNA codes that work perfectly well on a computer do not necessarily work in living, synthetically-engineered organisms. They also admit that the carefully-designed DNA codes may have unexpected side effects on an organism's behaviour. If this is the case, then the science we have to create must be a holistic one which takes into account all these factors and after-effects.

It follows that the likelihood of unexpected behaviours emanating from synthetic biology combinations is responsible for the fact that this science has developed no scientific methodology for testing the health and environmental safety implications of a new synthetic organism, apart from its recourse to the 'substantial equivalence' dogma, which enables the synthetic biologists to make 'a best guess' as to how the mixture of artificially-inserted genes and recipient organisms are likely to behave. This poses threats to human and plant life since the 'scientific' manipulations are based on guess-work. This is because the synthetic biologist who invents a synthetic microorganism cannot predict the effects of the release of such organism on human health and the environment with any degree of accuracy due to a lack of empirical evidence. There are greater ecological risks which are posed if such synthetic organisms are released or if they accidentally escape from biorefineries since they can outcross with natural species and contaminate microbial communities in the soil, seas and animals, including human beings [Ibid: 38]. Such a science cannot therefore be the basis of a new 'industrial revolution' that is being proposed and pursued.

ii. The new 'industrial revolution'

Be that as it may, this approach has, however, been the basis on which synthetic biology has been crafted to contribute to an 'industrial revolution' based on a new 'bio-economy'. This revolution, it is said, will depend on a mix of biomass feedstocks and new technologies, which are supposed to provide solutions to the current world's energy needs as well as solving the global food and environmental crises. The 'bio-economy' that is being touted describes the idea of an industrial order that relies on biological materials, processes and services instead of the old industrial 'raw materials' and labour services. Since many existing parts of the global economy are already biologically-based (agriculture, fishing, forestry), proponents of the 'new economy' often talk of a 'new bio-economy' to describe their particular re-invention of the global economy by merely clothing the current neoliberal economic and financial policies with new biological technologies and modes of production, without attempting to fundamentally problematise their basic economic assumptions and ideological claims behind the new strategies.

Thanks to these emerging technological changes, especially in the fields of nanotechnology and synthetic biology, biomass is being targeted by the 'new industry' as a source of living 'green' carbon to supplement or partially replace the 'black' fossil carbons of oil, coal and gas, which currently underpin the industrial economy. 'Swifts' are under way, to claim biomass as components of the new global industrial economy, that will draw its biomass resources from the countries that have been subjected to imperial rule for hundreds of years. This, in short, is a recolonisation of the former colonial countries along the lines already indicated by land grabs for biofuels and agrofuels. The new corporate drivers in this direction comprise forestry and agribusiness monopolies; high tech companies promoting biotech, nanotech and software; pharmaceutical, chemical and energy giants; financial institutions and investment banks; as well as consumer products and food companies.

There are four broad technological platforms that are being lined up to transform biomass into the new industrial revolution. These are, first, *combustion* techniques, which can burn extracts from biomass to the highest energy yield: open combustion with or without oxygen. This technique also includes biomass gasification, which entails burning extracts at very high temperatures with controlled amounts of oxygen as well as plasma arc gasification, which entails heating biomass with a high voltage of electric current. Secondly, there is the use of *chemistry*, which can be used to break down carbohydrates in biomass transformation into finer chemicals, polymers and other materials. For instance, thermochemical techniques can transform lignocellulosic material into hydrocarbons. Also, the extraction of proteins and amino acids yield valuable compounds as well as fermentation techniques, which are sometimes combined with genetic engineering and synthetic biology. These can produce proteins that can be refined further into plastics, fuels and chemicals. Chemistry is therefore one of the elements in the emerging Carbohydrates Economy – or the industrial use of plant materials – which will include the use of hemp in textiles, building materials, and industrial products. This will also involve the use of straw and other agricultural waste products to make building materials and the use of biochemicals to replace petrochemicals in a growing range of applications [Milani, 2000: 149].

The third platform is *biotechnology and genetic engineering*, which has been carried forward to be used for the fermentation of plant sugars as well as traditional plant breeding. Fermentation and breeding has been its speciality up to now and these processes have been used for thousands of years in a more organic manner for the same purpose. In ancient Egypt, for instance, a form of biotechnology was used that involved the application of living organisms in producing food and medicines. This was the case until experimenters discovered inadvertently the usefulness of one-celled organisms like yeast and bacteria. Yeast then began to be used for brewing beer as well as baking bread in ancient Egypt. The scientific study of the bio-chemical processes is less than 200 years old but current biotechnology has taken us to levels where life forms can be artificially created, which as we have seen, has reached a stage where it can be a threat to human life itself. Now new genetic engineering technologies have been introduced, which are being utilised to drive much of the industrial excitement around biomass. These include new approaches to genetic engineering (e.g. recombinant DNA) to modify plants to express more cellulose or to more readily break down elements for fermentation or to enable plant growth in less favourable soils and climatic conditions.

The fourth platform is *synthetic biology*, which has recently been added to the above platforms to produce novel organisms that are either more efficient at harvesting sunlight or nitrogen or that can generate entirely new enzymes (or biologically active proteins). These enzymes are used to carry chemical reactions or to produce new compounds from plant material. But synthetic biology has come into its own as 'the game changer' for biomass, as we have seen. It promises in the longer term to expand from the low-tech burning of biomass for electric production to the expansion of the chemical possibilities of turning biomass production into a global biomass grab for the new industrial bio-economy. In this 'revolutionary' role, synthetic biology will produce organisms with multiple traits from multiple organisms as we have described above. For instance, natural yeast has been routinely used by industry for years to behave like tiny bio-refineries in transforming cane sugar into ethanol or wheat into beer. But by altering the yeast (or other microbes), the same sugar feedstock under synthetic biology can be flexibly turned into novel products, depending on how the yeast's genetic information has been programmed.

This new synthetic biological technique can ingest sugar feedstocks and use them to excrete (or 'produce') hydrocarbon fuels with the properties of gasoline (instead of the usual ethanol) from billions of synthetic microbes contained in a single industrial vat. The same microbes, if differently programmed, can 'excrete' a polymer, a chemical to make synthetic rubber or a pharmaceutical product. In effect, the microbe has become, with synthetic biology, a production platform for different chemical compounds to make large chemical plants for industrial production. This indeed is the new biological engineering: whereby taking little genetic pieces of organism and 'programming' them, they can be put together into a whole 'industrial system'. In so doing, a cell can be designed to become a chemical factory for the future. This is the dream of synthetic biology; and a dream it is!

The new technological production platforms will be reinforced with *nanotechnology* with its well known suite of techniques of handling small and tiny substances. It will be used to manipulate the usual properties that substances exhibit when they are at the scale of atoms and molecules. Recently, there has developed an increasing industrial interest in transforming nano-scale structures found in biomass for industrial use. These include nanocellulose as a new commodity, which will take advantage of the long fibrous structure of cellulose to build new polymers, 'smarter' materials, nanosensors or even electronics. Research in nanobiotechnology will aim at modifying the nano-scale properties of living wood and other biomass feedstocks to alter their material or energy-producing properties into new industrial products.

This will lead to new products for new markets for nanomaterials, energy and pharmaceuticals as well as to the production of body armour, medical devices and food products. This transformation will also lead to the production of new forms of batteries, which have already been tried by nanoscientists from the University of Uppsala in Sweden who have developed high-quality paper batteries from coated cellulose fibres from hairy algae called *cladophora*. It is said that these batteries could hold 50-200 per cent more charge than the ordinary batteries and be recharged many hundreds of times faster than the conventional rechargeable batteries. One of the scientists has remarked: 'Try to imagine what we can create when a battery can be integrated into wallpapers, textiles, consumer packaging, diagnostic devices, etc. [Etc groups: 42].'

The promise of the new biomass bio-economy is to retool the same old industrial logic based on *private profit* onto the new system by switching from the old industrial products to the new products in the following areas but with the same profit motive, according to the ECT Group:

•   *Transport fuels:* Currently an estimated 70 per cent of petroleum ends up as liquid fuels for cars, trucks, airplanes and heating. Biofuels such as ethanol and biodiesel have marked the beginning of converting the liquid fuel market to biomass. A next generation of hydrocarbon biofuels is directly mimicking gasoline and jet fuel. From the short-lived corn ethanol boom of 2006-2008 to the new wave of venture capital and big oil companies sinking billions of dollars into biofuel start-ups, the biofuels industry is still regarded as a massive new source of revenue in an age of peak oil and carbon pricing. This is an area of great interest in terms of public policy as revealed by the World Bank internal memorandum, which was made public later to the effect that 73 per cent of the food price hikes of the 2006-2008 crisis was due to biofuels policies of the US and European governments. This prompted a massive switch away from wheat planting to rapeseed growing, coupled with a major diversion of corn and soy into ethanol and biodiesel production. But ethanol in particular proved to be a poor fuel in the production of fuel that produces less energy when combusted than gasoline. This new sector is being supported by government mandates for clean energy, stimulus funds and heavy investment by Big Oil.

•   *Electricity:* Coal, natural gas and petroleum are currently responsible for 67 per cent of global electricity production (International Energy

Agency, Key World Energy Statistics, 2008). However, co-firing of coal with biomass is on the increase and there is a growing move in many industrial cities to burn woodchips, vegetable oils, and municipal waste as the fuel for electricity production. Meanwhile, corporate interests are investigating ways to use nanocellulose and synthetic bacteria to make electric current from living cells, turning biomass to electricity without the need for turbines. A good example of investment in this sector is illustrated by biomass burning in the USA. The country generates one-third of all biomass electricity from this activity – making it the largest producer of biomass power in the world. As of October 2010, the grassroots group Energy Justice Network had mapped over 540 industrial power facilities burning biomass in the US; with a further 146 projected to be built. Currently, eight biomass power plants are connected to the electricity grid in twenty US states and currently generate about 10GW of power, which is half of the US renewable energy in an industry which is worth one billion dollars. Since 2000, biomass generation has risen by 25 per cent, according to the Biomass Power Association. But the environmental costs are too high in that the most straightforward impact of new biomass power facilities is the increased requirements for biomass, mainly wood, which is required 24-hours-a-day to keep the turbines turning. For example, the world's largest wood-burning biomass power generation station in Wales (currently under construction) aims to import over 3 billion tonnes of woodchips from the US, Canada, South America and Eastern Europe. According to a watchdog organisation, Biofuelwatch, 'the land area needed to grow this much biomass could be as large as one half-million hectares – ensuring the deforestation of an area three times the size of Liechtenstein every year' [Etc. Group, op. cit. 42].

• *Chemicals and plastics:* Currently around 10 per cent of global petroleum reserves are converted into plastics and petrochemicals. However, to hedge against rising petroleum prices, large chemical companies such as DuPont are setting ambitious targets to switch to supposedly renewable biomass feedstocks such as sugar for the production of bioplastics, textiles, fine and bulk chemicals. Making chemicals rather than transport fuels out of biomass is attractive because the markets are smaller and therefore easier to break into, while the prices of chemical products are on average two to four times higher. For this reason, venture capital investors are advising second-generation biofuel companies to branch out into chemicals as well as food as a secondary or even primary revenue stream. This is a field where synthetic biology is being deployed to make it possible to process and refine plant sugars within cells, instead of inside chemical factories, and therefore, more synthetic organisms are being fashioned to secrete chemicals that would previously have been refined from fossil sources. Now bio-based production is being applied across all sectors of chemical industry, including scents and flavourings, pharmaceuticals, bulk chemicals, fine and speciality chemicals, as well as polymers or plastics. Bioplastics is an area that is attracting a lot of investment excitement.

- *Fertiliser:* Global fertiliser production via the Haber Bosch process is an intensive user of natural gas. Proponents of biochar (carbonised biomass) claim that they have a bio-based replacement for improving soil fertility that could be produced on an industrial scale.

The proponents of the bio-economy are, however, faced with a strategic dilemma, which makes their claims contradictory and unachievable short of annihilating the universe. On the one hand, they claim that the mix of biomass feedstocks and new technologies will provide solutions to energy, food and environmental crises that are afflicting the global capitalist economy. However, the crises arise not out of a shortage of any of these commodities but out of organised scarcities of the commodities, which capitalism requires in creating profitable markets for them. The problem is the accumulated crises of the system caused by capitalist greed in search of private profit which has turned into a speculative activity against energy and food production, as we saw above. It is a dilemma because the new strategies that are being worked on come from the same monopolies that brought about the earlier crises. Therefore, if the new industries being created are to be based on the same profit motive, the same financial crises will occur, leading us to nowhere except alternative economic motives.

Their predictions that by 2050, the world population would have increased by 50 per cent and food demand by almost 100 per cent does not *ipso facto* mean that the new strategies they are proposing are the only ones that can solve these and related problems since the same approaches to the population and food issue through GMO production have never addressed the current food requirements. On the contrary, they have created shortages and scarcities which can enable them to engage in financial speculations to 'hedge' against risks of their valueless paper money. Their warning that climatic change will make, at the very worst, the situation worse is correct but the solution they propose will make global warming even worse. This is because they prescribe the same kind of solutions they have recommended in the past for agriculture. The recommendation has always been using more and more chemicals to rescue marginal lands and endangered habitats for crop production. Yet the same policymakers argue that the experimental technologies they are recommending will not only make everything alright, they will also mean imposing more demands on the soils and water supplies in the name of replacing the fossil carbons with living biomass energy. They are not able to provide a solution to this dilemma. According to the ECT group:

> 'If contained in biorefineries – despite the proliferation of production sites and the quantities involved – we are told there is little danger of environmental contamination and that these new biofactories can be fed sustainably. Those with similar hubris told us that nuclear power would be safe and too cheap to monitor; that the chemical age would end hunger and disease; that biotechnology would end hunger and disease, too – and not contaminate; and – only recently – that climate change is probably a figment of our imagination. In other words, gamble with Gaia ... using experimental life forms on the back of contested hypotheses. More than a biomass grab or a Land Grab, this is an Earth Grab [Ibid: 55].'

This Earth grab will be a culmination of the activities of the transnational corporations that have for two centuries been working feverishly to own and control the entire earth for private profit of a few capitalists against the majority of humankind. They have advocated the conservation of resources only when they served to enhance their future access to those resources for profit and not because they regard such resources as a common heritage of humankind. For instance, in 1958 Laurance Rockefeller, a member of the Rockefeller family and the Rockefeller Foundation, which has worked so hard to promote eugenics, genetics, population and food control, formed the Conservation Foundation, which was to complement his brother, John's, Population Council. According to Engdahl:

'Both the Population Council and the Conservation Foundation were united around the unspoken theme that natural resources must be conserved, but conserved from use by smaller businesses and individuals, in order that select global corporations should be able to claim them, thus establishing a kind of strategic denial policy masquerading as conservation [Engdahl, op. cit. 93].'

This was part of a strategy of population control and family planning in order to stop the 'over-population' of the earth by poor, 'inferior' races. It was, in the words of Engdahl: 'preparing a global assault on "inferior peoples", under the name of "choice", of family planning and of averting the danger of "over-population" – a myth their think-tanks and publicity machine produced to convince ordinary citizens of the urgency of their goals' [Ibid: 93]. As we have seen, these goals led to the pursuance of tied 'food aid', which compelled poor countries in the Third World to accept US foreign policies approved by the transnational agribusiness and other corporations. Genetic engineering is connected with these policies, including the latest synthetic biology policies.

In short, the biomass-based bio-economy that is being touted as the 'new' industry is a continuation of these institutions that 'conserve' the world's natural resources only when they can be exploited by the same corporations that are controlled by rich families. They are the ones who look down on the rest of the world's population as 'poor', 'backward' and 'inferior'. This is because the same transnational companies who fostered dependence on the petroleum economy during the twentieth century are the same companies now telling the world about the new bio-economy to 'save' the world from existing crises in the twenty-first century, which they had created. Indeed, the invasion of biomass on the scale proposed is an invasion of the lands mainly occupied by the 'poor' and 'backward' populations in the global South. Much of this biomass, which encompasses over 230 billion tons of living matter that the earth produces every year, such as trees, bushes, grasses, algae, crops and microbes are located there. This annual bounty, also known as the earth's 'primary production', is most abundantly found in the tropical oceans, forests and fast-growing grasslands where it sustains the livelihoods, cultures and basic needs of most of the world's indigenous communities. Currently, only one quarter of land-based biomass for basic needs and industrial production is being used – and hardly any oceanic biomass – leaving over 90 per cent of the planet's full biomass production still untouched by the industrial economy.

This is the natural bounty that the new biotechnological sciences, particularly in the fields of nanotechnology and synthetic biology, are threatening to invade in the name of constructing a new 'industrial revolution' based on exploiting nature to its logical conclusion. This stock of the annually produced biomass is being targeted by the genetic and synthetic agricultural industry as a source of living 'green' carbon to replace or supplement the supplies of 'black' fossil carbons of oil, coal and gas that currently underpin industrial economies of the world. From generating electricity to producing fuels, fertilisers and chemicals, 'swifts' are already being made to purportedly elevate the importance of biomass as a critical component in the new global industrial economy.

But the matter is not as it is presented. The situation is far from being a benign and beneficial 'switch' from black carbon fossil energy to green carbon. It is a brutal and real grab of land and other natural resources from the poor populations by the dominant economic interests in the world economy, which is in crisis. Their declared objective is to capture biomass as a new source of wealth for their exploitation. Plundering the biomass of the people of the world in order to pass on the costs of such an economy to the poor people of the world – and the use of their natural resources – is what is left for the current economic and social system to cheaply run the remainder of the capitalist industrial economies. Such a drive is a new form of twenty-first century imperialism that is bound to deepen inequality as well as exacerbate existing poverty, hunger, disease and other social ills afflicting poor communities of the world. Liquidating fragile ecosystems for their carbon and sugar stocks is also a suicidal move on an already overstressed planet that is overheating in a global climatic system. Therefore, we have not only to dismiss the false promises being made of a new clean and green bio-economy but interrogate and resist seriously their inflated claims, which are intended to seduce the world into accepting the latest assault on land, nature and the livelihoods of the people of the world, which will result in annihilating all living organisms, including human life itself, from the face of the universe.

We need to develop a 'glocal' cross-social movement across the globe to create a coalition of social forces to engage in deeper conversation aimed at saving the universe for all humanity and not for a few corporations of the rich who claim the world's resources. This deeper glocal conversation must move towards the mainstreaming of the *Precautionary Principle*, which states that: 'when an activity raises threats of harm to the environment or human health, precautionary measures should be taken, even if some cause and effect relationships are not fully established scientifically'. This principle was arrived at as a comprehensive response to environmental degradation caused by the actions of agribusiness by a group of activists, scientists, lawyers, policymakers, and environmentalists at Wingspread, headquarters of the Johnson Foundation in Racine, Wisconsin, USA. They were advocating for the emergence of a glocal strategy to defend the ecosystem.

The Glocal Coalition of Social Forces for Another World must work towards the 2012 Rio+20 Summit Agenda 21 to encourage a full glocal debate on all the socio-economic, environmental and ethical implications of the synthetic biology bio-economy and its relationship to the biomass use as well as the application of synthetic biology and the governance of the emerging technologies in general.

The UN System's Environmental Management Group (EMG) should undertake a major study of the implications of the new bio-economy, particularly as it relates to the livelihoods, biodiversity, and the rights of the affected communities throughout the world. This study must engage all governments, civil society organisations, the widest range of concerned parties, and forest and farming indigenous communities themselves. These studies and consultations must lead to a legally-binding International Treaty for the Evaluation of New Technologies (ICENT), which would allow for the close monitoring of the harmful effects of the new technologies by governments and all the affected peoples and organisations.

Already the International Movement for Ecological Agriculture in a declaration issued in Penang, Malaysia in 1990, had called for a similar strategy after it observed that the Green Revolution in Asia had 'singularly failed to address the primary causes of hunger', which were mainly due to 'a history of unjust social and economic systems which, in combination with ecological degradation, had marginalised the poor and deprived them of the means to eat'. The Declaration had called 'A radically different approach ... one that seeks the regeneration of ecosystems through ecological agriculture, and which brings about the wider social, economic and political changes necessary to ensure food security and social justice for all'. These declarations reveal that a lot of research work has already been undertaken that demonstrates that the continuation of the Green Revolution through the Genetic and Synthetic Revolution can only lead to a downward destructive spiral of the ecosystem, which must be stopped in favour of a regenerative agriculture, if the small farmer, the backbone of agriculture and food security, is to be saved.

[G] GLOCAL POLITICAL IMPLICATIONS OF THE AGRICULTURAL CRISIS

The developments outlined above have demonstrated the 'success' that global capitalism has scored in its 'modernisation' and 'development' paradigms throughout the world with its positive and dire consequences. It has engulfed the entire globe and turned it into a single territory controlled by an alliance of a global Superclass in all countries of the world. This Superclass comprises the global political and economic elites who are no longer tied to a particular territory called 'nation', since globalisation has for long eroded what used to be referred to as 'national sovereignty'. This is what neoliberalism in the phase of late globalised finance capitalism has achieved for global capitalism.

William J. Robinson and Jerry Harris [2000] remind us that the entire process central to capitalist globalisation has been the emergence of a transnational class formation, 'which has proceeded step by step with the internationalisation of capital and the global integration of national productive structures'. They add:

'Given the transnational integration of national economies, the mobility of capital and the global fragmentation and decentralisation of accumulation circuits, class formation is progressively less tied to territoriality [Ibid: 12].

The two authors point out that the Transnational Capitalist Class – the Superclass – which has emerged is a global ruling class. This class, they add, has no borders and is composed of the technocrats, media, corporate banking, social and political elites of the world. It is a ruling class because it controls the levers of an emergent transnational state apparatus and global decision-making, which takes place at many levels. Their class enemy is no longer the working class, which they have destroyed through deindustrialisation of the economy turning the former 'middle classes' into the new working class, which is also becoming increasingly unemployed and declassed. Indeed, this declassing group of new workers is now seen as the more dangerous in that being increasingly unemployed it can mobilise and cause revolutionary situations to erupt throughout the world to which it is linked electronically. It is therefore on the line for elimination at the hands of the Superclass, if it can.

The other social forces that are seen as problematic for the new ruling Superclass are the peasant classes, who are now regarded as having the power over land and the natural resources of the world in some regions, especially the indigenous peoples. They are, in fact, the real class that now claims sovereignty for the small farmers for food rights and human rights. By eliminating the middle class the Superclass will target these lower classes of the world that were poverty-stricken even prior to the capitalist crisis. These lower classes will suffer the greatest distress from the capitalist crisis, 'most probably leading to a massive reduction in population levels', which is already happening with the massive land grabbings and intensification of wars and conflicts over resources such as diamonds and minerals of different kinds. This is taking place and is continuing in a number of African countries such as in the Democratic Republic of Congo, Rwanda, Burundi – including genocides and ethnic cleansings of various kinds and outright starvation in many regions of the world.

This globally impoverished class, led mainly by the declassed 'middle class' and poor peasantry and indigenous communities, is also seen as dangerous because of its capacity to resist the implications of the failed capitalist post-colonial States. As we shall see below, the transnational agribusinesses are, after the 2008 global capitalist crisis, engaged in land grabbing, especially in Africa, where it is said that the 'marginal lands' are still in abundance. But the peasant classes on these lands are already resisting the grabbings and that is why the new ruling Superclass would regard them as enemies to be fought in their plans to create a new global economic order based on synthetic biology. Peasant and middle class uprisings such as those which have taken place in Tunisia, Egypt and other Arab States are still current and can be re-ignited in these and other marginalised countries. The Arab Spring is becoming a global resistance movement. We have witnessed the still on-going Zapatista Movement and the emergent indigenous political and social movements coming up in these regions of South America, Central America and Asia.

Furthermore, the attempt to localise politics and economies is on the increase everywhere. New communities, as well as new 'commons' and cooperative movements are emerging or are being formed by the marginalised groups with the support of the declassed middle classes. Even in the capitalist world, thousands of local economic, social and political movements have emerged leading to anti-globalisation movements being formed and waging struggles against globalisation. This is a 'living democracy' in which direct democracy is replacing elite 'representative' democracy and where 'free money' credit is replacing commodity-tied money and credit, leading to solidarity economies, which are increasingly turning into 'green economies', expanding locally and globally into new *glocal movements* and challenging the prevailing global capitalist profit economy, as alternative systems [von Werlhof, 2010: 116-136]. This is a favourable environment in which to put forward a *glocal circular green economy movement* – not only in theory but also in practise – through new forms of production and holistic knowledge systems.

It is therefore significant that there is an emerging convergence movement of scientists that is drawing attention to the need to reconnect with nature and restore the ecosystem. This movement is beginning to connect with the movement of indigenous communities. According to one of the scientists promoting this new approach, the new science has begun to focus on addressing the adverse impacts on nature by modern mainstream scientific activities. This is intended to restore the framework which makes us, as living beings, dependent on parts of nature and which simultaneously makes nature the object of our thoughts and actions. In this situation scientists can no longer confront the universe as outside and 'objective' observers since science must now recognise the participation of man with the universe and vice-versa. This reconvergence will give, once more, cosmic significance to the human being [Anshen, R. T., 1986: xi-xxix]. There has been a line of African scholars and scientists such as Prof Cheickh Anta Diop who have been advocating a similar approach.

A new epistemology for a new form of glocal circular green economy requires a complete revamping of the contemporary scientific paradigms towards a new world-view that accommodates nature. The *Agriculture at a Crossroads* report, referred to above, has already indicated the need for a broader approach to the agricultural economy. The interdisciplinary team of researchers who researched on this report from a number of inter-governmental organisations called for the combination of various forms of exogenous scientific knowledge in the natural, agronomic, economic, and social sciences to become integrated in the search for a new paradigm for agriculture. These modern forms of knowledge should, in their view, be combined with the highly diverse forms of 'local', 'traditional' or 'endogenous' forms of knowledge in order to rejuvenate agricultural production. They pointed out that these different forms of knowledge are represented by different local actors such as farmers, traders, and craftsmen as well as external actor groups such as civil servants, extensionists, researchers, service providers, etc. They represent different kinds of knowledge, which can be referred to as 'knowledge systems' that need to be recognised as such. Combining endogenous and exogenous knowledge, they point out, can be achieved by increased participation of 'end users' – including marginalised and poor actors – in the different forms of research and development. The report adds that while the initial focus of combining knowledge is on increasing participation at local levels, today emphasis is also shifting towards the *upscaling* of participatory processes into the meso- and macro-levels of social organisation resulting in multilevel and multistakeholder approaches to learning and doing things.

The researchers further pointed out that when taking into account the centrality and value of endogenous, traditional or local forms of knowledge related to agricultural development, such as through ethnological approaches in sciences studying agricultural soils, plants and animals, it is necessary to reflect on the ethical and epistemological implications related to the integration of different knowledge systems. This is because the integration of, and cooperation between, different knowledge systems is often hampered by interactions that do not take into account the need for processes of communication to move beyond the practical and generally tangible technological, economic, ecological and social effects of innovations. In the long run, they point out, innovation can only be successful if it 'makes sense' to all the parties involved, and hence innovation needs to be integrated into (and by) the different knowledge systems involved into a holistic system. This is also particularly important for innovations in rural development, they add.

There is growing consensus among researchers concerned with sustainable agriculture that no single group of actors should appropriate the right to themselves define what type of combinations should exist between scientific and local forms of knowledge. As a consequence, participatory forms of *co-production of knowledge*, based on *social learning* among actors involved, should become a key feature of sustainable agriculture and resource management. This means that the role of science within a process of participatory knowledge production must be redefined in new ways. Instead of striving to find the ultimate truth through scientific research, the scientific community must complement conventional and generally discipline-based knowledge production with inter- and transdisciplinary approaches to knowledge production.

The particularity of a transdisciplinary approach to research and learning is that it implies examining 'real-world problems' from a perspective that firstly goes beyond single disciplines by combining natural, technical, economic and social sciences, and secondly by including the poorest people in society as another dimension to such integrated knowledge processes. For instance, the *Crossroads* report adds that much private and public research and development is spent on crops consumed by the rich and the middle classes such as corn, wheat, maize and rice, while very little attention is devoted to research on 'poor man's crops' such as cassava, millet, sorghum and potatoes. Furthermore, the report points out, it has not proved easy for researchers and extension workers to adapt their established practises to the new way of understanding rural development as part of a broader approach based on co-produced knowledge by all actors involved in production.

According to the Report, many public research and development (R&D) bodies of National Agricultural Research Systems (NARS) are finding it difficult to deal with poor farmer and peasant economy-based issues in many countries. The problems range from resource constraints on the one hand to rigid, disciplinary-bound research paradigms on the other. Often there is a lack of engagement with client sectors and unwillingness to exchange and co-generate knowledge with other research bodies in the sector. This is also related to the process of identifying research problems, which is often based solely on perceptions of disciplinary-based researchers with incentive systems usually grounded mainly on the number of publications by individual scholars in accredited journals.

The inevitable result is that all too often resource allocation to the NARS does not pay off in terms of economic, social and environmental development possibilities for poor farmers. While a number of countries have initiated some remedial policies on these issues, the relevant literature shows that there is still some way to go to overcome this divide. The difficulties of more equality-based engagement with farmers, peasants, or 'clients', according to the Report, has also to do with an understanding of the reasons guiding rural actors' decisions, actions and livelihoods that is too narrowly defined.

Furthermore, traditionally the passing on of the results of agricultural research to users has hitherto been handled by state-funded extension services. Not only have these institutions suffered through World Bank and

IMF structural adjustment programmes in Africa, the increasing number of questions have also been raised by the extension systems themselves as operational and organisational mechanisms for handling such research results. There is also evidence of an increased need to engage in partnerships in order to reconceptualise (in theory and practise) the delivery of technology in the context of a broader system of agricultural knowledge, science and technology that is based on the paradigms of knowledge co-produced by scientists, policymakers and community groups.

These new partners should include private sector organisations, non-governmental organisations (NGOs), community-based organisations (CBOs) and civil society organisations (CSOs) and movements that are able to bring skills and knowledge to bear because of the close relationships they have established with specific communities. Therefore, today's challenges in community development in poor countries make it more imperative and compelling for institutions of higher education to create effective changes of vision and prepare professionals to lead innovative rural development processes alongside the farming rural communities. Training, capacity building, and the reinforcement of small-scale farmers' skills to enable them to participate in the agriculture supply chains are urgently required if improvements in agricultural production on a new ecological basis are to take place. A new approach in conceptualising transdisciplinary approaches to research was developed by the interdisciplinary researcher team and has been summarised by them as shown below. This model of knowledge co-production can contribute to a new system aimed at creating a new understanding of agricultural knowledge, science and technology (AKST).

Towards a New Epistemology of Agricology

The above model of AKST demonstrates the need for an integrated system that can address the challenges facing agriculture in a post-capitalist crisis world. But they are solutions that can take hold only when the direct producers and producers on the ground are firm on what their programme is. They must clarify their local or traditional knowledge and innovation ideas and practises within their knowledge systems before they can effectively be real co-producers with the other partners. The response to these multifaceted crises must first address the problems that have been generated within the industrial agricultural economy and go beyond current practises and theories that are being advanced within the mainstream knowledge systems –however progressive they may appear – because this is not the first time these new ideas have been advanced.

As we have seen above, the central issue is to move from the current agricultural knowledge systems and practises to a more broad-based community-responsive system of eco-agriculture that addresses not only the needs of self-sustenance, energy, and health of the present generations but also the needs of future generations. But this demand must be built within the spiritual and moral systems of the indigenous knowledge as indeed they are often built in order to ensure the continuity of the humanistic considerations in those spiritual-moral systems for human survival and holistic existence. The responses must address the need for the restoration of the ecosystem, especially the soils, water systems and the environment. These restorations cannot take place unless our spiritual relationships with nature are redefined and re-established. We cannot begin from a vacuum but must build on an accumulated record of humanistic heritages that need to be rediscovered and reactivated in a new environment.

We have to build on what we have achieved to move to greater heights.

CONCEPTUAL FRAMEWORK OF THE REPORT ON AGRICULTURE, KNOWLEDGE, SCIENCE AND TECHNOLOGY

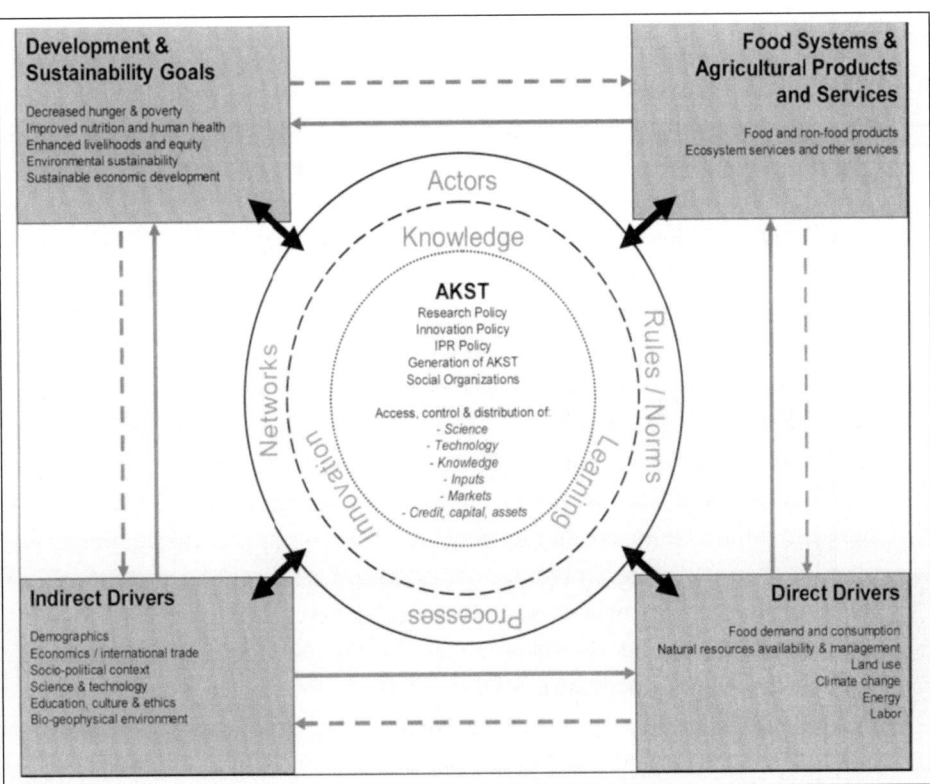

Source: *Agriculture at a Crossroads* report

In this regard, we have to examine what exists within the existing [A] Indigenous knowledge systems and farmer innovation; [B] Master the divide between the rural and the urban economy; [C] Examine the resilience of ecosystems that still exist in small farming activities; [D] Re-conceptualise the circular ecosystems; [E] Re-establish traditional governance and justice; Move from agriculture to agricology.

[A] Indigenous Knowledge Systems and Farmer Innovation

i. General

The World Bank defines indigenous knowledge (IK) as local knowledge, which is unique to every culture or society. IK is the basis for local-level decision making in a variety of knowledge fields that cuts across all disciplines and beyond such as agriculture, medicine, education, philosophy, history, mineralogy, animal husbandry, natural-resource management, spirituality and other activities in communities. IK provides problem-solving strategies for communities on a day-to-day basis and although it includes what are called sciences and humanities and social sciences, it does not distinguish between them. In that sense, IKS can be referred to as truly 'transdisciplinary'. It is collective knowledge held by communities rather than by individuals in tacit form and therefore it is difficult to codify into modern mathematical-logico languages. This is why it is difficult, if not impossible, to integrate it with other knowledge systems. This is because it is embedded in community practises, institutions, relationships and rituals as well as in prophecies and divinations and shamanism. IK provides problem-solving strategies for local communities, especially for the poor in society who are all practitioners and contributors to IKS; and IKS is therefore highly hermeneutic and dialogical. This is why the oral form in which it is embedded and transmitted is open-ended and accommodative of new ideas and inspirations even from the dead in the form of spirits.

Thus the *Agriculture at a Crossroads* report is right in insisting that it is necessary to completely revamp the contemporary scientific paradigms in order to create a new world-view that accommodates nature; and that is why the report speaks of combining endogenous and exogenous knowledge with other knowledge systems rather than integrating them. They argue that this can only be achieved by increasing the participation of real end users who must include marginalised and poor actors in the different forms of research and development. The report also refers to a further shifting towards *upscaling* participatory processes into the 'meso- and macro-levels of social organisation' resulting in multilevel and multistakeholder approaches to learning and doing things together. This, they conclude, is a movement to a new epistemology for a new form of agriculture, which we refer to here as *agricology*. IK is therefore an important contribution to global development knowledge, although it is an underutilised resource in the development process and transformation process.

In order to make IKS an equal partner in the collection and combining of knowledge systems, it is necessary to investigate and learn afresh how local communities generate, utilise and perpetuate IKS through philosophising, theorising and practicing it. This is why IKS is highly innovative. This is also why a number of mainstream institutions such as the World Bank have tried to co-opt IKS and put it to their use for purposes of implementing their programmes. In an introductory paper, the Bank notes that IKS is a key element of the social capital of the poor communities which constitutes their main asset in the efforts to gain control of their own lives. It then advises that the potential contribution of IKS to locally managed, sustainable, and cost-effective survival strategies should be promoted in the development process.

In order to facilitate the integration of IKS into mainstream operations, the African Department of the World Bank launched the Indigenous Knowledge for Development Program in 1998. The programme was intended to assist communities and governments to integrate indigenous knowledge into the development programmes it had approved. But this attempt was not in fact an integral part of the programmes of the Bank. It was just attached to its activities in order to give the impression that it is sensitive to the knowledge of indigenous peoples without any serious consideration of how such knowledge is generated, philosophised, theorised, and practised by these communities before it could recommend it to be integrated in its development programmes. In its other programmes, the Bank has worked towards the modernisation of traditional systems into 'modern' and 'rational' systems. This is contradictory to what it advocates in its IKS policy and it is dangerous to talk of 'integrating' IKS to other, modern systems of knowledge without those systems being first revamped.

In this section, we want to demonstrate how IK is in fact being rediscovered and how it is now being used by the communities and its sympathisers from the mainstream to challenge some of the approaches that the mainstream scientific knowledge systems have promoted and which have brought about disasters that could have been averted if all knowledge systems were accessed and utilised. This is why those systems must first be revamped by interrogating them and exposing their erroneous basis. What the experiences from the grassroots communities show is that IKS is in fact very central to the knowledge systems of the world and has therefore an innovative potential in every direction, including science. We shall examine those potentialities in this section in order to demonstrate how IKS has to be recognised as a partner to all other systems of knowledge which humanity should exploit in order to save Mother Earth from destruction and extinction.

The *Agriculture at a Crossroads* report already referred to has recently defined innovation to include activities in indigenous communities. It pointed out that innovation is a *network of agents* usually organised in interdisciplinary and trans-disciplinary teams, with interactions that determine the relevant innovative impact of knowledge interventions, including those associated with scientific research. This concept is now used as a kind of shorthand for the network of inter-organisational linkages that successful countries have developed as a support system for economic production. In this connection it has been explicitly recognised that economic creativity actually relies on the quality of 'technology linkages' and 'knowledge flows' amongst and between different economic agents and not from one source.

The report adds that some approaches in this direction suggest that innovation systems cannot be separated from the social, political and cultural context from which they emerge. This implies the need to focus on those factors that enable the emergence of the 'innovative potentials', rather than on factors related directly to specific innovations. Therefore, the circular green economy (agricology) epistemology can be enhanced more by finding out what exists on the ground and what has been preserved from ancient times by indigenous communities. This knowledge from the ancient wisdom of Osiris and Isis has continuously been passed on by word of mouth as a living knowledge. This requires recognising the spiritual basis, which has enabled this continuous process to go on until the present.

This means that approaches based on *linear* systems of research-to-extension-to-application have to be revamped and replaced by approaches that focus on processes of communication, mutual deliberation, interaction and doing: that is to say, iterative collective learning and action based on experience and new innovative knowledge. More concretely, the report adds, this implies that sustainable use of natural resources requires a shift from a focus on technological and organisational innovation to a focus on the norms, rules and values under which such innovation takes place, including IKS approaches. In short, there must be a face-to-face interrogation between IKS and the linear scientific approaches to arrive at a representative general approach.

This new enhanced model of collaboration holds that the values, rules, norms and practises that are relevant to the promotion of a new agriculture must be constantly produced and reproduced by the real social actors who are embedded in the social networks and organisations to which they belong. Social networks, in this sense, are important spaces where the actors involved in the *co-production* of knowledge share, exchange, compare and eventually socialise their collectively- and individually-realised perceptions of what is important, good, and bad; and foreground the visions they have for their own families, communities, and wider social groups to which they belong.

ii. How IKS is generated and innovated

Indigenous Knowledge Systems or Traditional Knowledge Systems are in general low-input agricultural systems which are based on extensive and applied knowledge about natural processes which preserve life and the animal world. They have enabled millions of people to subsist for thousands of years in some of the most hostile environments all over the world. However, many traditional agricultural knowledge systems have fallen into decline due to neglect, under-utilisation and under-estimation by the western 'science-based' analytical approaches to knowledge production. Globalising market economies, commercialising agriculture with the introduction of export crops and Green Revolution technologies and the intellectual property protection of seeds by agribusinesses have been shown above to be factors contributing to this decline and under-utilisation. This, however, has not meant that the rural communities have abandoned this form of knowledge because it is obsolete: in fact, the original agricultural systems, although weakened and sidelined, have remained dynamic in many parts of the world and continue to be the basis of much innovation in agriculture. Communities continue to manage agricultural genetic diversity effectively through experimentation with traditional on-farm and modern crop varieties. They continue to produce their own products whose performances in many cases are better than those produced by modern agricultural extension services and not at all 'substantially equivalent' to GMO products.

There has emerged a positive awareness in a number of countries about the importance of indigenous agricultural practises in the innovation of the agricultural sector. A number of scholars from different Non-Governmental Organisations (NGOs) have come together to challenge the existing mainstream system of scientific research and innovation based on reductionism and have attempted to link up directly with the farmers as innovating agents. They have argued that current research strategies are commodity-focused when promoting agricultural development, as has been evident in the strategies for the Green Revolution based on the Transfer of Technology Model. They point out that these strategies have had limited impact on the intended beneficiaries, as the complexity of their livelihood and farming systems is ignored in the reductionist 'scientific' research strategies.

The shift to embrace farmer innovation systems is evident by the application of participatory approaches to research and development such as on-farm experimentation through participatory research, participatory technology and innovation development. This approach builds on farmers' local innovations, local resources and indigenous knowledge. Decision-making is left in the hands of the farmers themselves who, in consultation with other research and development actors, play a facilitating role in collaborations. These experiences also reveal that there are complexities in these relationships in terms of the diverse researchers addressing the needs and priorities of their constituencies and stakeholders at the local levels. Therefore, these practitioners insist that the different researchers must balance their requirements for scientific rigour with the community's need to address real-life, time-bound problems.

In the case of IKS, the promotion of farmer innovation in the main fosters individuals and/or groups to discover and develop better ways of managing resources by building on and expanding the boundaries of their existing indigenous knowledge through *doing, using, and interacting.* In this way, innovations occur both in technical and socio-institutional frameworks in an organic way. In these collaborative research arrangements, other researchers argue that innovations must be broadly related to the introduction, adoption, or creation of either or both elements of 'new knowledge', by which they mean ideas, skills or experiences; and that these require 'new organisation' in the form of principles, forms, networks or mechanisms as part of the innovation process. This is why complexities in the relationships occur.

For example, the concept of farmer innovation is therefore applied to agricultural technology processes to include new knowledge such as 'sustainable biotechnology' that aims at improving rural livelihoods for sustainable development while ensuring inter-institutional and farmer learning. The advocates of such new thinking argue that farmer innovators are those farmers and/or land users who innovate, test and try new methods of conservation or production, on their own initiative, often using ideas from various sources. But these 'sources' are not always 'neutral' for they have their interests built within the 'new technologies' and innovations. Therefore, rural innovators tend to be curious, creative, proud of their innovations, willing to take risks but at the same time cautious. They can be skilful in blending their own ideas with ideas picked up elsewhere but they are also able to reject those they disapprove.

In the mainstream reductionist research, scientists, with very little involvement of the consumers and producers, largely decide the research agenda without any consultations. In fact the research process should be demand-led, which builds on farmer innovations and problem-posing, rather than the other way around. It is a task that requires changes in the mainstream organisational culture, structures, institutions and systems. Learning from research collaboration in order to implement such changes is a major challenge facing the agriculture in crisis, since innovations and technical progress must be the result of a complex set of relationships among actors, as we have argued above.

The driving forces behind the emergence of the farmer-innovation paradigms arise from a number of factors such as the constraints to crop and animal production due to pest and disease infestation and soil fertility depletion to which agricultural production is highly sensitive. These driving forces also attribute the paradigm shift to increased population pressure while the natural resource base remains constant, as well as to weak interactions among stakeholders. Economic globalisation and market incentives, through liberalised trade, and climate change, are also singled out for the change in the attitude to farmer innovation. Innovations are, therefore, in this view, needed to meet the needs, standards and quality, among others, for the diverse markets and needs that exist both internally and externally.

Promoting Local Innovations (PROLINNOVA) practitioners (an

organisation supported by the World Bank) insist that the key ingredients for livelihood improvement are not external inputs but are to be found in the labour, knowledge and local management capacities that enable people to manipulate skilfully the local resources for their own benefits. They point out that most rural development efforts have failed to mobilise and enhance these 'internal inputs', which are crucial to any *national systems of innovation*. The farmer innovation system approach allows for interactions and integration between different stakeholders, resulting in *social learning*. This enables the stakeholders to identify and recognise their experimentation efforts, responsibilities, strengths and weaknesses, thereby strengthening participation and community innovation processes.

As we shall see below, farmer innovation system approaches have been adapted to integrated soil and water conservation, integrated pest management (IPM) and Farmer Field School (FFS) approaches, among others. Consequently, there is a higher adaptation of technologies by farmers than is usually supposed. In a co-production mode, conversely, the other research actors also learn from the farmers about their farming systems, and about the actual constraints and potentials of the communities. This interaction is crucial to the future management of eco-based agriculture and is the basis for the demand that all knowledge systems be recognised and 'combined' in order to effectively deal with the emerging issues in agriculture.

iii. Some Case Studies

But there is more that is done by the small farmers by way of innovation than is generally recognised. Utilising 'rural appraisal' approaches, two South African researchers, Tim Hart and Johann Mouton [2005], carried out research on small-scale farmers in certain villages, which produced interesting results. These scholars also pointed out during their research that recent developments had shown an increased awareness about the failure of the application of conventional agricultural practises to the different types of geographical zones in which modern agriculture was practised. This resulted in a greater attention being paid to local or indigenous knowledge to address the gaps that had been created. Their case study focused on the indigenous knowledge relating to the cultivation and use of traditional vegetables in a rural parish in western Uganda, using a participatory research method called *Rapid Rural Appraisal*.

The results of the study illustrated the importance of understanding indigenous knowledge for future agricultural research and extension activities. The results also discovered a number of important issues regarding the mainstream understanding of indigenous knowledge, namely that it is often contrasted with conventional agricultural practises, being influenced by its purposes and the resources to which it has access. To place some of the results of their research in a broader context, a comparison was made with similar studies done in other African countries.

From these comparisons, the two researchers came to the conclusion that a greater understanding of the utilisation of appropriate indigenous knowledge would contribute to the success of future agricultural interventions in development policies. They observed that agricultural development projects in Africa have predominantly followed the input-output transfer of technology development model, which assumed that a country's technological, economic and social development could only be induced by external technological interventions in order to 'catch-up' with the developed world. They observed that projects based on these models typically identify beneficiaries who receive a range of inputs that are expected to bring about 'development' but which do not achieve the objectives.

Based on research experiences in one parish in south-western Uganda, the study argued that the indigenous or local knowledge appeared to be important in achieving sustainable agriculture, especially regarding food security, and that it was therefore important to future agricultural development programmes and strategies to rely on this approach. The researchers visited these resource-poor and marginal areas and discovered that the farmers survived and, in some cases, prospered by almost exclusively relying on local resources, farming within the parameters set by their access to these limited resources. The study highlighted a number of issues relating to the current mainstream understanding of indigenous knowledge, which the mainstream arrogantly regarded as erroneous and irrelevant to the communities.

The two researchers pointed out that in Africa crops identified as traditional vegetables are important to food security and are typically consumed in conjunction with the common staples of rice, maize, millet and highland cooking bananas (also known as 'Matoke') as well as with meat on occasions. These 'vegetables' also served as food security, which is attributed to the fact that the vegetables grow easily; requiring less external inputs such as mechanisation and agro-chemicals. The farmers also adapt cultivation practises to locally available resources and ecological conditions. The vegetables were also found to be more nutritious and cheaper to purchase in urban area markets than exotic vegetables. The researchers also observed that the current cultivation of these crops by the small famers was in line with sustainable agriculture approaches and the production was continually adapted to meet changing circumstances.

The two researchers noted that in Uganda, crops labelled as 'traditional vegetables' were becoming increasingly important as foodstuffs for both rural and urban residents, following increased trends in urbanisation. While official research and extension services had generally ignored these crops, local producers and many consumers believed that these vegetables were more nutritious than exotic vegetables such as carrots, cabbages and lettuces. They noted that there was a lack of official interest in these crops and therefore concluded that local producers were the custodians of most of the knowledge pertaining to their cultivation and use. This, according to the researchers, makes it imperative that the local people must participate in any project or programme involving traditional vegetable crops and food security programmes proposed by the government.

The two researchers also noted that there was evidence that the majority of rural inhabitants in Uganda suffered from nutrition deficiency due to the lack of adequate feeding in most areas of the country. They called on all scientists involved in agriculture to collaborate in a non-official way with the rural famers so that more research on traditional vegetables and food crops can be done in a collaborative manner since the vegetable crops were becoming increasingly important in terms of food security in some areas. It was believed that such collaborative work would lead to a greater awareness of the benefits and use of these crops in a sustainable manner. But these efforts are highly disregarded by lack of governmental support in the form of legislation to protect IK knowledge systems.

That is why a number of legal scholars [Cottier and Panizzon, 2004] have called for the devising of a new form of intellectual property rights (IPR) protection regime that recognises the social value of traditional knowledge and promotes its 'integration' into the domestic and international trade regimes while respecting and preserving local autonomy and cultural values. This is prompted by the goal of promoting the social, economic, and ecological development of rural areas in conformity with the needs of the rural communities and the market. According to the two authors:

'It responds to concerns about fairness and equity in international economic relations affecting the livelihoods of the bulk of the world's population. The topic is also of importance in the context of redefining the relationship between public goods, private rights, and the transfer of technology. Taken together, these concerns lead us to evaluate the policies and legal instruments that are best suited to achieve equity, validation, and sustainability, while preserving open access to plant genetic materials for scientific research [Ibid: 371].'

The two legal researchers also pointed out that the sole reason why traditional knowledge was currently not protected was that it did not, according to the conventional view, 'yield innovations'. The authors, however, pointed out that by redefining public goods and private rights we shall be able to recognise, for instance, the farmer's right to the information that their seeds carry when the agribusinesses undertake genetic engineering about the nature of seeds they want to 'engineer'. By so doing, they turn these engineered 'inventions' into their private property when the information in the seeds they extract from belongs to indigenous communities in the form of traditional knowledge. They proposed an intellectual property regime for traditional knowledge that aims at not only rewarding the intellectual efforts of the community as embodied in knowledge about seeds, it also ensures that the allocation of seeds would remain under the sole proprietorship of that community instead of it being left entirely to the public domain 'in what still amounts to an informal sector of a subsistence economy in many countries' [Ibid: 373]. This attitude should change and biotechnology should be introduced to rural communities in such packages that build on their already acquired knowledge and not in a manner that is 'engineered' to benefit transnational businesses that are accountable to no one but their investors and executives.

The two authors further propose that farmer's intellectual property rights should be recognised as a *collective property* rather than individual rights. They point out that the argument that collective rights are not recognised under international law is false. They draw attention to the fact that many of the existing internationally protected intellectual property rights (IPR) in the form of patents and trademarks are of the nature of collective rights, which are recognised in international law. They further point out that trademark law protects collective marks: 'Even the enforcement of copyrights and related rights has a long tradition of operating on the basis of collection societies to which authors and artists belong.' According to them, 'formed under private law these associations may easily cross national boundaries and comprise producers and owners unrelated in territorial terms' [Ibid: 385].

The two legal researchers therefore recommend the recognition of a Traditional Intellectual Property Rights regime (TIP Rights) in international law as a legitimate exercise on behalf of these communities that would enable them to trade in their products and carry out innovations which are protected by both domestic and international law. This is the way to go. This proves that IKS is, unlike all other systems of knowledge, a core knowledge system and is more than entitled to be recognised as a core participant in co-production with other systems of knowledge. In fact it deserves to be referred to as the *Mother Knowledge System*, from which other systems draw meaning and can strengthen themselves. IKS is also entitled to be known as the *Original Knowledge System*, imbued with divine origins deep into the Cradle of Humanity and has a continuous linkage to our contemporary knowledge and Ways of Knowing.

B: MASTERING THE RURAL AND THE URBAN DIVIDE

One of the fundamental changes brought about by modernity and its introduction of the industrial economy and its overwhelming of agriculture into its segment was the increasing divide that this economy introduced between the rural and the urban economies. Historically, it is the population and the resources from the rural areas that made it possible for the urban economy to emerge as a 'modern' sector. In chronological terms, it is the Enclosure Movement in Europe that led to the 'modernisation' of agriculture in Europe, which also formed the basis for the 'primitive accumulation' of capital that formed the foundation for a revolutionary capitalist economy. In Africa and the rest of the non-European World, it was colonisation that created the colonial enclaves that later became the towns and the cities.

In terms of the agricultural transformation through the conversion of common lands to private landed property, colonisation created real conditions for the emergence of the rural and the urban economies. The urban was the space created for the foreigner and coloniser, while the rural countryside was for the natives. In short, the urban emerged as territory for exploiting the rural but also as an outpost for organising colonial production to service the 'Home Market'. The exploitation of the rural peasantry and their land for raw materials and food products for the industries of Europe meant that the rural became a locus for cheap labour that was exploited to produce 'cheap' products for the empire. The 'subsistence' that was said to remain was an economy that catered for the natives and at the same time cross-subsidised the wages of the rural labourers who were recruited to work in the urban areas and mines. This explains why wages paid to these workers were kept low for colonial capital to make super-profits since part of their livelihood could be provided by the families.

The rural-urban divide therefore became essential for the promotion of 'modernity' and the 'modernisation' of the 'traditional' society into a new 'progressive' society. One of the ways of advancing the new society was to transform the traditional agricultural and pastoral economies into colonial enclave economies. The introduction of the Green Revolution was one of the means of hastening this process; but as we have seen it also consolidated the rural-urban divide. The dispossession of traditional peasant lands for large-scale agriculture, and later for the application of chemicals to agricultural production, displaced large numbers of peasant small famers, many of whom drifted to the urban areas to become impoverished as the 'urban poor'. This was especially so in India where this process went deep as the Green Revolution took shape. The decline caused by the new industrial agriculture in the colonies and the use of chemical fertilisers impoverished not only the soils and the water supply, it also impoverished the small farmers to become the urban unemployed.

This form of agriculture was also adopted by the socialist world in their emphasis on the mechanised production of State farms and agricultural collectives and cooperatives. But the failure of this model in the socialist countries gives an example of how the rural-urban divide could be overcome. As an economic strategy, the 'Stalinist command economy' resorted to the use of chemicals in order to 'catch up' and 'bypass' the West and improve the livelihoods of the peasant farmers. The ill-effects of this model were especially felt in Cuba, which had adopted the Soviet Model of agricultural and industrial development. For a period, from the time Cuba adopted the Soviet Model, Cuba relied on oil imports from the USSR for its agriculture and industry. Before 1989, however, Cuba was a model for a 'socialist' Green Revolution farm economy, based on large State-owned farms, which depended on large quantities of imported oil, chemicals and machinery from the USSR. Cuba produces crops with high-yield variety seeds, which it exported to the USSR to pay for its imports from there.

Under commercial agreements signed with the USSR, Cuba was an oil-driven economy with 98 per cent of its oil needs coming from Russia. In 1988, it imported 12-13 million tons of Soviet oil from which Cuba re-exported two million tons to other countries such as Angola. As the USSR undertook its own economic reforms, it reduced the amount of oil which it had exported to Cuba, and by 1989 Cuba was forced to cut its oil imports to 10 million tons from the promised 13 million. By the end of 1991, only six million tons could be imported to Cuba and the shortfall began to affect its economy greatly. This in turn adversely affected Cuba's exports to the USSR. As a result, with reduced imports of inputs and food supplies from the Soviet Bloc, Cuba was forced to shut down some of its factories, just as food security was also affected and the technology base eroded. The combined withdrawal of Soviet support and the US trade embargo, soon exposed the Green Revolution Model Cuba had adopted and this exposed it to its worst food crisis in history [Murphy, 2000].

The food crisis, shortage of medicines and fuel shortages in the rural areas, in particular, fuelled the rural-urban migration to the cities, especially Havana. From 16 541 people in 1994, the numbers migrating grew to 28 193 by 1996. Despite the introduction of police controls the influxes continued until the city was populated to an average of 3000 people per square kilometre. Now 74 per cent of the population was urban-based, which increased the pressures on food production with only half of the previous chemical and oil inputs coming from the USSR. Although the population was able to be fed as well as previously, the country was not importing any food supplies from the USSR or former Socialist countries. The country was forced to adopt self-reliance as its economic strategy. This entailed in the case of agriculture a policy of paying farmers higher crop prices in order to induce them to produce. The self-reliance approach also forced a reliance on agro-ecological technology, smaller production units, and the promotion of urban agriculture.

The most important step in the promotion of self-reliance was the restructuring of agriculture in two ways: from a large-scale, high-input monoculture system to a smaller-scale, organic and semi-organic farming system, and then from a rural to a rural-urban based system. This was a revolutionary step in a double sense. It was a reversal of the previous Stalinist Command socialist economy into a kind of mixed system where the smaller producer was to play a big role. It was also a reversal of the general trend in population movement in the modern world from a shift to cities, which relied on food supplies from the countryside, to a new agriculture that was both rural- and urban-based at the same time.

The new focus was on using low-cost and environmentally safe inputs in agriculture as well as relocating production closer to consumption in order to reduce costs of transportation given shortages of fuel. Urban-based agriculture was the key to the adoption of this strategy. The strategy was not state-sponsored as such but was in part a spontaneous response and decentralised movement that had begun to take shape in the cities. The people in both the rural and urban areas also responded enthusiastically to the policy initiative that the government had begun after 1993. By 1994, more than 8000 city farms had been created in Havana alone. Schools and offices began to produce their own food supplies by using whatever space was available, including the front lawns of municipal buildings for growing mainly vegetables. There was also a change in the way people employed themselves, in that the urban gardens were farmed by retirees and semi-retirees, which increased the level of new employment for old age. Also urban women began to play a larger role in urban farms than their rural counterparts were doing.

By 1998, an estimated 541 000 tons of food were being produced in Havana alone for local consumption. Food quality also improved with a greater variety of vegetables and fruits being grown. Urban gardens continued to grow in other cities and some urban areas were growing as much as 30 per cent of their food needs. The State encouraged the people in the urban areas to utilise unused and semi-used urban and semi-urban lands for food production. Urban planning laws were changed to allow this to take place. The state also permitted the opening of local farmers markets, thereby legalising direct sales from farmers to consumers. Unlike the old days when everything had to be sold through state structures, this development opened up some 'mixed economy' practises that encouraged small farmer activities. Prices were also deregulated so that it was now possible for urban-based farmers to make two or three times the income of rural professionals. Seeds were produced and sold within the cities. Compost and biofertilisers were also produced and sold within the urban areas by control agents at low cost.

The State research institutions had now to carry out research that reinforced the new economy. New biological and organic gardening techniques were developed. Organic alternatives to chemical controls were put in place enabling urban farming to become completely organic. A new law was passed prohibiting the use of any chemicals and pesticides for agricultural purposes anywhere within the city limits. This spurred agricultural production that was encouraged by the new diversified market-based system of marketing, which was now available to farmers. The United Nations Food and Agricultural Organisation estimated that between 1994 and 1998, Cuba had tripled the production of tubers and plantains, as well as doubled the production of vegetables, which doubled again in 1999. Potatoes increased from 188 000 tonnes in 1994 to 330 000 tonnes in 1998, while beans increased by 60 per cent and citrus by 110 per cent from 1994 to 1999. In the countryside, food production had also increased and by 1999, many city dwellers were trekking back to the rural areas. This was the first time in history for this to occur on this scale in any part of the world.

Here in the countryside, agro-ecological methods had already been introduced, largely out of the necessity of coping without chemical fertilisers. This was supported by the State, through research and fundamental policy shifts at the highest levels of government. Agro-ecological farming in the countryside and organic farming in the cities were the two strategies that stabilised both urban and rural populations. The agro-ecological methods introduced in the rural areas included: locally-produced biopesticide substitutes; complex agrosystems designed to take advantage of ecological interactions and synergisms between biotic and abiotic factors that enhanced soil fertility; biological pest control; and the achieving of higher productivity through internal processes. Other factors were those also undertaken in Asia as we have seen. These included: the recycling of nutrients and biomass within the system; the addition of organic matter to improve soil quality and active soil biology; soil and water conservation; the diversification of agrosystems in time and space; the integration of crops and livestock; and the integration of farm components to increase biological efficiencies and preserve productive capacity.

These developments compelled the Cuban government to unveil a major reorganisation of agriculture in the country by restructuring state farms into private cooperatives, called Basic Units of Cooperative Production (UBPC). This restructuring arose out of the new perception that smaller farms were better easily managed and better able to take on sustainable agricultural practises. Under the new system, the State continued to own land but it leased it on a long-term basis, rent-free to the farmers. The cooperative, and not the State, owned the production and the members' earnings were based on their share of the cooperative income. The UBPC also owned the buildings and farm equipment, which were purchased from the government at discount prices with long-term, low-interest loans. Most of the cooperatives produce sugar according to quotas, limiting any other crops they might produce. Therefore, they had less to sell in the agricultural markets, and this restricted their incomes and options. Thus the breakup of the large state farms enabled the freed land to be turned into private farms as well as agricultural cooperatives.

Agricultural Production Cooperatives (CPAs), which had been created in the old system to attain greater production, marketing and economic efficiency and which were in decline, also began to rebound in the 1990s and became the models for the UBPCs, except that the farmers in the CPAs were able to own their land. There were also Credit and Service Cooperatives (CCS) in association with small landowners joining to receive credit and services from the state agencies through this means. Through them, they were able to share machinery and equipment, and thereby take advantage of economies of scale. CCS members purchased inputs and sold products at fixed prices through state agencies, based on production plans and contracts established with the State distribution system. Any production above the agreed and beyond the contracted quantities was sold at farmer's free market prices. Hence, CCS farmers had higher incomes than members of other cooperatives. Through these means, they are able to triple or quadruple their incomes.

Urban production was not restricted to Havana alone but was allowed spread to all major urban areas throughout the country, as we noted above. This expansion turned the urban areas to become part of the rural economy and vice-versa. From Santiago de Cuba in the East to Pinar del Rio in the West thousands of urban gardens began to blossom. In these areas, some 300 000 Cubans were growing their own fruits and vegetables and selling the surplus to their neighbours. However, although the urban agriculture was totally organic, the country as a whole was not. There were still pockets where chemicals were still being used, but these were drastically reduced. Before the crisis of 1989, Cuba used more than one million tons of synthetic fertilisers a year. By 2000, it was using about 90 000 tons in total. In regard to herbicides and pesticides, consumption was reduced from 35 000 tons to only 1000 tons by this time, indicating the extent to which organic farming and agro-ecological agriculture had taken root in the economy of the country.

Thus, Cuba's contribution and example to the rest of the world has been its achievements in developing urban agriculture on a sustainable basis. According to Cuba's Ministry of Agriculture, some 150 000 acres of land is being cultivated in urban and suburban settings in thousands of community farms ranging from modest courtyards to production sites that fill entire city blocks. *Organponicos*, as they are called, show how a combination of grassroots efforts and official support can result in sweeping changes. It also shows how neighbourhoods can combine their efforts and available resources to feed themselves, contrary to Malthusian pessimistic predictions of starvations. From *ad hoc* emergency structures, the *Organponicos* have transformed Cuban agriculture with the support of the Cuban government stepping in to provide key infrastructural support and assisting with information dissemination and skills sharing. According to Mae-Wan Ho:

'Most *organponicos* are built on land unsuitable for cultivation; they rely on raised planter beds. Once the organponicos are laid out, the work remains labour-intensive. All planting and weeding is done by hand, as is harvesting. Soil fertility is maintained by worm composting. Farms feed their excess biomass, along with manure from nearby rural farms, to worms that produce nutrient-rich fertilisers. Crews spread about two pounds of compost per square yard on the bed tops before each new planting [Mae-Wan-Ho, 2009: 29].'

The *Organponicos* are self-organised and self-managed by the owners. There is no boss supervising others, and each person seems to understand well their roles and what is expected of them. The work occurs fluidly, 'with a quiet grace' [Ibid.]. Gardeners come from all walks of life: doctors, artists, teachers, etc. What makes the *organponicos* work so well is the existence of the *hybrid public-private partnership*, which appears to work well. In return for providing the land, the State receives a portion of the produce, usually about one-fifth or 20 per cent of the product, which is used at state-run day care centres, schools and hospitals. The workers keep the rest, which they sell at produce stands located right at the farm. The city of Havana now produces enough food for each resident to receive a daily serving of 280 grams of fruit and vegetables daily. Wolfe observes that:

'Urban agriculture nationwide reduces the dependence of the urban populations on the rural produce. Apart from *organponicos*, there are over 104 000 small plots, patios and popular gardens, very small parcels of land covering an area of over 3600 ha, producing more than the *organponicos* combined. There are also self-provisioning farms around factories, offices and businesses, more than 300 in Havana alone ... Shaded cultivation and apartment-style production allow year-round cultivation when the sun is at its most intense. Cultivation is also done with diverse soil substrates and nutrient solutions, mini-planting beds, small containers, balconies, roofs, etc., with minimal use of soil. Produce levels of vegetables have doubled and tripled every year since 1994, and urban gardens now produce about 60 per cent of all vegetables consumed in Cuba, but only 50 per cent of all vegetables consumed in Havana. ...There is so much for the world to learn from the Cuban experience, not least of which agriculture without fossil fuels is not only possible but also highly productive and health-promoting in more ways than one [Ibid.].'

Wikipedia, the free encyclopedia, points out that the garden city philosophy is not new but a method of urban planning that was initiated in 1898 by Sir Ebenezer Howard in the United Kingdom. Garden cities were intended to be planned, self-contained communities which were surrounded by 'greenbelts' or parks, containing proportionate areas of residences, industry, and agriculture. This was laid out in the 1898 book by Howard: *To-morrow: a Peaceful Path to Real Reform*. The diagram below illustrates some of his idealised garden city plan, which was intended to house 32 000 people on a site of 6000 acres (2400 ha), planned in a concentric pattern with open spaces, public parks and six radial boulevards, 120 ft (37 m) wide, extending from the centre. The garden city was to be self-sufficient and when it had a full population, another garden city would be developed nearby. Howard envisaged a cluster of several garden cities as satellites of a central city of 50 000 people, linked by road and rail.

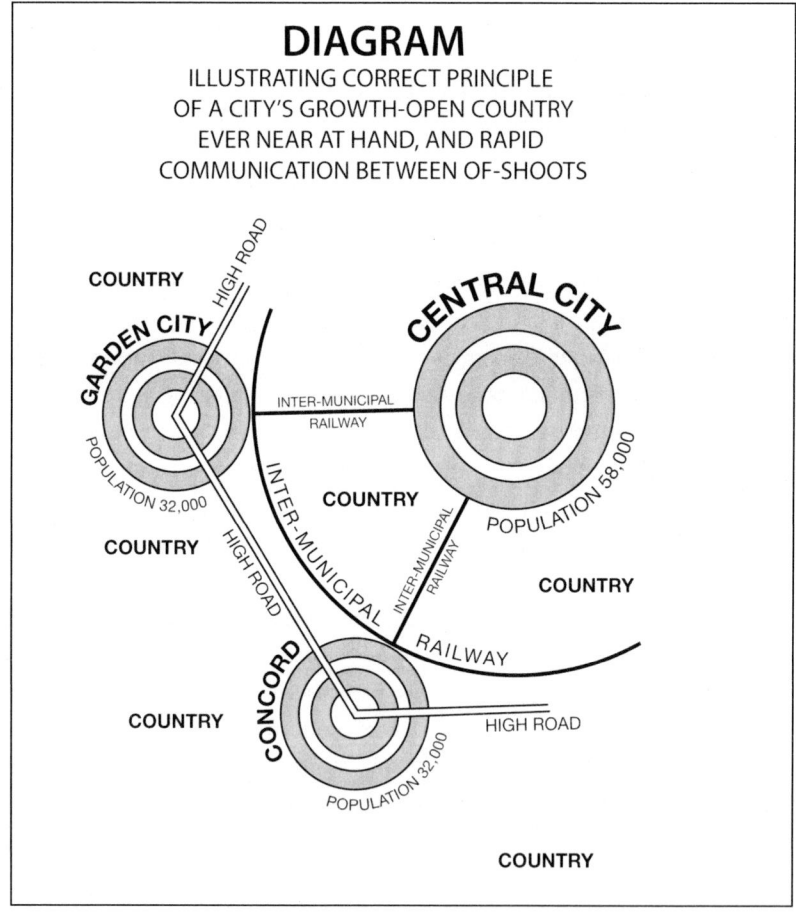

These models are now being replicated in a number of urban areas in the developed capitalist world. 'Green cities', 'sustainable cities' or 'green municipal networks' have emerged to adopt some of the principles of linking former rural areas to urban areas. In the UK at least 20 such projects around cities and towns have been developed. These achievements demonstrate that it is possible for cities with an industrial heritage to overcome the legacy of the past and perform well as new sustainable green cities aiming at retrofit programmes, ecological restoration projects, community gardening and community-supported agriculture. According to Milani, such projects can bring in the local state and local utilities to be involved in such programmes. The administration of such projects can later devolve to neighbourhood activists and local voluntary experts: 'In this sense, (successful) green economic development would absorb the State into the community in a positive sense. According to Milani, these organic activities could provide a base for fully-fledged green municipal political structures for the future [Milani, 2000: 190].'

Indeed, he argues that post-industrial agricultural transformation will represent a major decentralisation of food production, which will take the primary form of a *gardening revolution*. The transformation would see an entire city becoming a kind of garden, attempting to tap the latent permaculture productivity of back alleys, rooftops, walls, sidewalks, parks, industrial lands, plazas, and more. In his view, edible landscaping would be a big part of the multi-functioning use of plants: 'which could also help purify the air, soil, and water; shape outdoor space; and provide air conditioning, shade, windbreaks, food and habitat for wildlife; a productive sink for organic wastes; and even industrial feedstocks' [Ibid: 105]. Because a green economy is mainly a closed-loop system, food production would be an obvious way of dealing with the volumes of organic waste that a city produces. According to Milani:

'This would mean not only composting food scraps and yard waste but also putting human waste to better use than polluting the water. Like renewable energy, modern compositing toilet technology is safe and cost-effective but suffers from cultural and institutional barriers. An explosion of urban food production would be likely the biggest factor spurring more ecological handling of organic waste. Support for more conventional organic farms is also, of course, essential in creating regenerative food systems. Besides providing healthy food for city people, they can help redefine the city/country boundaries and roll back sub-urban sprawl [Ibid.].'

Milani and this group of green economists argue that Green Cities would also be consistent with eco-manufacturing, which must move away from petrochemical-based production toward plant-based materials, the so-called Carbohydrate-Economy. In their view: 'This shift could have a tremendous revitalising impact on rural communities, but even Green Cities could provide a certain amount of feedstocks for local industry, with manufacturing increasingly linked to horticulture [Ibid.].' They point out that urban-ecological food networks in many cities are already combining activities on various levels of the food system such as education about food, nutrition, and ecology; training in organic gardening, farming, and permaculture; support for community-supported agriculture and support for rooftop gardening and community gardens; as well as special organic food programmes for the poor and unemployed [Ibid: 106]. This is the only way humanity can avoid the disastrous consequences of the 'new industrial revolution' being proposed by 'synthetic biology'.

[C] THE RESILIENCE OF ECOSYSTEMS AND THE SMALL FARMING COMMUNITIES

The persistence of the small farmer in the modern industrial economy is indeed a good thing for the emergence of a new society based on agricology. Despite all the pressures of the capitalist economy, the small farmer has managed to sustain himself and his family in many countries of the world. The spirit of the small farmers' struggles for survival against all odds is the same spirit that strengthened the farming family unit to preserve the environment despite all the obstacles they have encountered with the pressures of industrial agriculture. This is sometimes referred to as 'the spirit of the land' that keeps the peasant farmers going on the basis of inherited knowledge from their ancestors through the 'living word'. Sometimes the peasants implore the ancestors to come to their aid to sustain their efforts when they are faced with calamities such as when droughts occur and there is a food crisis. This is how the indigenous knowledge systems are maintained and perpetuated, as we saw above. It is indeed a spiritual endeavour.

Therefore, for the small famer working under conditions of traditional agriculture, there can be no separation between the body, the spirit, and the mind as the Cartesian 'scientific' epistemology would want us to believe. The three entities constitute a holistic interlinked existence, which is partly physical, metaphysical and spiritual. This unity of the trio is what is behind African philosophies and belief systems. So when the African peasants speak of the land 'belonging' to three 'beings' – the living, the dead and the unborn – they speak of a spiritual and physical unity in being human. Of the three beings, one is a physical being in the form of *the living* – who are in existence – the other two are non-physical metaphysical beings, which the African philosopher, Professor Ramose, has called 'the ontology of invisible beings', or African metaphysics [Ramose, 2002].

These three beings, Prof Ramose argues, may not be seen or known to exist, but the fact that they are unknown does not mean that they are unknowable or unbelievable. The Africans, in this understanding, according to Ramose *believe* in the existence and beingness of the unknown, which has a direct influence on their own beingness. It is this existence of the invisible beings, he adds, that is at the base of *Ubuntu* metaphysics and which is behind the belief in the supernatural; and this belief plays a role in African demands for reconciliation and harmony. This, according to Prof Ramose, explains why *Ubuntu* philosophy and religion have no separate and specific theologies. Through these invisible forces Africans seek explanations to certain happenings, which cannot otherwise be explained by 'normal' or 'rational', 'scientific' means.

Therefore, African politics and law based on *Ubuntu* is a unity, which arises out of the recognition of the continuous oneness and wholeness of *the living, the living-dead and the unborn*. Their continuous being is expressed in the 'living word', through which knowledge and language are preserved and perpetually communicated. The combination of politics and laws is an embodiness of rules of behaviour, which are expressed in the flows of daily life. It is for this reason that African political philosophy responds easily and organically to the demands for reconciliation as a means of restoring the equilibrium of the flow of life when it is disturbed. This is what the resilience of the peasant life is all about in their relationships with the dead, the unborn and nature in the pursuance of their survival on this planet. This is a restorative philosophy, which aims at re-enacting the balance in relations not only between human beings but also between human beings and nature as well as the animal world. The restorative philosophy itself is an expression of the resilience of all life and a belief in the continuity of the universe of which humanity is part.

This resilience is expressed in the failure of the modern economy to sustain all life forms. The attempt to re-assert indigenous knowledge systems is therefore an attempt to overcome these failures of modernity and global capitalism. It manifests itself as resilience and this is why IKS is a core knowledge system as we concluded above. It combines all the knowledge the humans experience on earth in their struggles for existence and this is best done by the peasant farmer. This is why 'resilience' has attracted attention on the part of scholars and governments as a 'subject of transdisciplinary research'. As a new 'discipline', resilience is said to provide 'important insights' into how society can organise in times of crisis to prevent further distress [Hoffmaister, 2009: 18]. But this is only one aspect of the IKS, which is over-encompassing; which cannot be understood in isolation of the other practises of IKS.

According to Hoffmaister, the concept 'resilience' is beginning to resonate 'more and more' not only in academic circles, but also in international negotiations, where government officials are recognising the 'increasing resilience' of small farmer responses to many of the challenges facing our planet, especially to the crisis in the food systems [Ibid.]. But this recognition of 'resilience' is not coming out of the 'ideas' or heads and brains of academics and government officials from nowhere. They are merely referring the resilience of the peasants on the ground and the experiences of the small farmers involved in the persistence of their existence. In short, they are talking about the knowledge of the peasant farmers, which is embodied in IKS.

It is true, as Hoffmaister points out, that 'agro-ecological approaches can provide strategies to confront challenges, such as drought and desertification while enhancing food security', but where are these 'agro-ecological approaches' to come from? If scholars and government officials have been engaged in 'transdisciplinary researches' to discover what lies behind the 'resilience' of the small farmer survival activities, they must be aware that these activities exist on the ground where the small farmers exist. It is this existence and the knowledge that sustains them that enables the researchers to 'design' their research 'projects' in the first place. Without these experiences the researchers have no basis for even contemplating 'transdisciplinary approaches'. It is possible to think of a 'transdisciplinary approach' because the resilient activities being investigated cannot fall under the auspices of a single academic system. Therefore, the idea of trans-disciplinarity comes from the small famers' epistemologies and cosmologies.

Hoffmaister correctly recognises that: 'the concept (of resilience) is only relevant when (it is) put in context and not (when it is seen) as an empty word in the development discourse'. Hence his two questions: 'resilience for what?' and 'resilience for whom?' Hoffmaister, for instance, hypothesises that the resilience of food systems in the drylands of Africa 'may depend on the ability of (small) farmers to manage the land with appropriate agro-ecological practises to withstand perturbations', which allows them to recover and adapt to change. This, he points out, can only happen when these farmers are able to create 'institutions that support their coping strategies, rather than being passive targets of technical fixes' [Ibid: 18].

This is a correct observation and is attested to by the experience of small farmers from Niger, West Africa, where the struggle against desertification has been under way for some time in fighting the effects of global climatic change. Hoffmaister quotes the World Resource Institute as reporting that in the drylands of Niger, 'communities have managed to reverse trends of desertification and drought through agro-ecological management while enhancing their resilience and increasing their food security'. The report continues that after the devastating droughts of the 1980s, small farmers and 'other actors' were able to reverse the drought through 'farmer-managed natural regeneration' and by 'adopting simple, low-cost techniques for managing the natural regeneration of trees and shrubs and the use of soil and water conservation programmes' [WRI, 2008]. The 'other actors' seem to have been the Greening Movement of elite practitioners that worked with the farmers. However, it is the structures that the small farmers had put on the ground that enabled this 'farmer-managed regeneration' to take place and hence their resilience in the face of an advancing desert.

But food security requires more complex interlinkages involving multilevel and decentralised institutions to enhance information flow as well as combining different types of knowledge for learning. This is what provides a forum for different levels of actors to collaborate during periods of reorganisation prior to and after the crises. In the case of Niger, the restorative process was assisted by the government establishing a secretariat of the *1993 Rural Code*, in collaboration with NGOs and other actors. This code helped to create the environment for collaboration, which enabled the different actors to support farming practises rather than dictating to the farmers, which would not have worked in the first place. The farmers were able to enhance their resilience not through adopting expensive technologies, but through utilising farmer-led agro-ecological practises, such as rock-lining, improved traditional planting pits, etc.

This was in fact a socio-ecological process that the farmers alone could implement. In this way, the government became an ally of the farmers and not a dictator of what they ought to do. The state modelled its laws on the farmers' socio-ecological processes and this is what was used as the experience for writing the *Rural Code*, though they assisted the farmers' self-managed activities. The Code provided farmers with rights and a framework which enabled them to manage more effectively the natural regeneration approaches in collaboration with agricultural extension services that were supportive to the process. The enhanced resilience in the region arose from these approaches that increased the capacity of the farmers to cope with drought. This was because the farmers were able to practise natural regeneration aimed at stocking grains during good years and harvesting trees for food and income and in that way obtaining better insulation against cyclical droughts [WRI, 2008].

According to Hoffmaister, the experience of Niger serves to highlight that 'resilience' is not a concept that stands alone, but is an approach that can advance how farming communities can organise against shocks and crises. The question posed by him as to what resilience is for, and for whom, has enabled the questioning of the modern paradigms and epistemologies of technology-driven food systems and to support practises that make socio-ecological sense instead. Interest groups including scholars and NGOs, according to him, have used the recent food crisis to sell solutions, when according to him 'solutions have always been with us' among the small farmers [Ibid: 20]. Enhancing soil quality by reversing drought in Niger 'has required preserving ecosystem functions that may not seem to contribute directly to food production', but that have other benefits for the system. Ecosystems such as food provisioning, 'depend on the maintenance of biological integrity and diversity in agro-ecological systems [Altieri, 1999]. Therefore, it is not a 'market' question as to who decides what to grow and who are the farmers. According to Hoffmaister:

'Enhancing the resilience of food systems requires strengthening from within, learning and using knowledge and resources to accommodate to change, and seizing the opportunities to enhance ecosystems, as farmers have done in Niger [Ibid: 20].'

Thus it is only by increasing the resilience within the agro-ecosystems, that organic agriculture can increase its ability to continue functioning when faced with the unexpected events such as climate change. This is closely linked to farm biodiversity, because practises that enhance biodiversity allow farms to regenerate natural ecological processes, enabling them to better respond to change and reduce risk. Therefore, organic farming practises that preserve soil fertility and maintain or increase organic matter can reduce the negative effects of drought while increasing the productivity of food crops. In addition, water-harvesting practises allow farmers to rely on stored water during droughts. Other practises such as crop residue retention, mulching and agro-forestry, conserve soil moisture and protect crops against microclimate extremes. Indigenous and traditional knowledge systems are key sources of information on adaptive capacity, centred on the selective, experimental and resilient capabilities of farmers. These innovative activities of the farmers have enabled the rural communities to become resilient when faced with the global climatic change and this in turn has enhanced the resilience of the ecosystem. The two resiliencies have depended on each other.

[D] Re-conceptualising the Circular Ecosystems

The above experiences have shown how important it is to re-conceptualise our relations with nature and the ecosystem; and the experience has also demonstrated that we cannot do this without the cooperation and adopting of the knowledge systems of the small farmers and their epistemologies. Attempts must be made to return to indigenous modes of knowledge and practises that have been tested in the past and which have been found to be innovative, scientific and resilient and combine them with new knowledge systems.

Immediate steps must be taken to promote local food systems towards circular green economies, known to sustain livelihoods at meso and micro levels. These have been challenged by globalised food systems, which as we saw above, have been drawn into capitalist speculative financing activities, thereby undermining the productive capacities of farmers and the natural systems. This trend brings opportunities to new social classes (such as capitalists and middle classes) but it also threatens the livelihoods, sovereignties and resources of marginalised communities and indigenous peoples. In some countries, social, ethical, and cultural values have been successfully integrated in commercial mechanisms of agricultural farming but this has been selective, to meet the needs of finance capital. Fair trade and ethnic labelling are examples of institutional options that have been considered by those who wish to promote effective measures to protect the interests of the marginalised communities and revitalise rural livelihoods and food cultures. But these measures are limited in scope.

According to the *Agriculture at a Crossroads* report, production systems dominated by export markets are weakened by erratic changes in international markets and have sparked growing concerns about the sustainability of long-distance food shipping and the ecological footprints and social impacts of these practises. Local consumption and domestic as well as regional outlets for farmers' products can alleviate the risks inherent in international trade (formerly 'distance trade'). Awareness about the importance of ensuring full and meaningful participation of multiple stakeholders in international and public sector agricultural knowledge and science policy formation has increased, and so changes can be made. These policies have focused on the multi-functionality of agriculture as a system beyond industrial agriculture. The number and diversity of actors engaged in the management of agricultural resources such as germplasm has declined over time. This trend reduces options for responding to uncertainties of the future, particularly of poor communities spread around the world. It increases asymmetries in access to germplasm and increases the vulnerabilities of the poor and indigenous communities.

According to the report, participatory plant breeding provides strong evidence that diverse actors can be engaged in an effective practise for achieving and sustaining broader goals of sustainability and development by bringing together the skills and techniques of advanced and conventional breeding and farmers' preferences and germplasm management capacities and skills, including seed production for sale. Further development and expansion would require adjustment of varietal release protocols and appropriate policy recognition under the International Union for the Protection of New Varieties of Plants (UPOV).

According to the Report, the debates surrounding the use of synthetic pesticides have led to new arrangements that have increased the awareness, availability and effectiveness of the range of options for pest management. Institutional responses have included the strengthening of regulatory controls over synthetic chemical pesticides at global and national levels; growing consumer and retail markets for pesticide-free and organic products; the removal of highly toxic products from sale; the development of less acutely toxic products; and more precise means of delivery and education of users in safe and sustainable practises. What constitutes safe and sustainable practise has been redefined in widely varying ways by different actors reflecting different conditions of use, as well as different assessments of acceptable tradeoffs. The availability of and capacity to assess, compare and choose from a wide range of options in pest management is critical to strengthening small farmers' ability to incorporate effective strategies that are safe, sustainable and effective in actual conditions of use.

Integrated Pest Management (IPM) exemplifies a flexible and wide-reaching arrangement of actors, institutions and practises that better addresses the needs of high-value crops. This has enabled large surpluses of a narrow range of basic grains and protein foods to be generated, traded and also moved relatively quickly to meet emergency and humanitarian needs. It has eased hunger and reduced poverty as well as kept food prices stable and low relative to other prices and allowed investment in other economic sectors.

However, according to the Report, the ecological and cultural context of farming is always and necessarily 'situated' and cannot – unlike functions such as water use or carbon trading – be physically exchanged. Advances, especially in the ecological sciences and socioeconomic research, as well as drivers originating in civil society movements, have mobilised science, knowledge and technology in support of approaches appreciative of place-specific, multidimensional and multifunctional opportunities in communities. Examples include trading arrangements connecting those willing to pay for specific ecological values and those who manage the resources that are valued; urban councils using rate levies to pay farmers for the maintenance of surrounding recreational green space or for ecosystem services such as spreading flood water on their fields; hydroelectric companies paying farmers to practise conservation tillage to avoid silting behind the dams and improve communal water supplies; farmers' markets; and community-supported agriculture.

The resulting flows of products and services are embedded in a web of institutional arrangements and relationships at varying scales, such as farmers' organisations, industrial districts, commodity chains, *terroirs*, production areas, natural resource management areas, ethnic territories, administrative divisions, nations, and global trading networks. Farmers are simultaneously members of a variety of institutions and relationships that frame their opportunities and constraints, offering incentives and penalties that are sometimes contradictory. Farmers need strategic ability and time to select and interpret the relevant information constituted in these institutions and relationships. The various ways of organising science, knowledge and technology over the last 60 years have taken different approaches to farmers' strategic roles but these needs should be improved. These new arrangements and relationships constitute part of the reconceptualisation of the new approaches to ecoagriculture and other forms of farming.

According to the *Agriculture at a Crossroads* report, the concept of 'sustainability', although controversial and diffuse due to existing conflicting definitions and interpretations of its meaning, is nevertheless a useful beginning point because it captures a set of concerns about agriculture which is conceived as the result of the co-evolution of socioeconomic and natural systems. A wider current understanding of the agricultural context requires the study between agriculture and the global environment, as well as the cultural and social systems, given the fact that agricultural development is the result of a complex interaction of a multitude of factors that are inseparable. It is through this deeper and interconnected holistic understanding of the ecology of agricultural systems and their actors that will open up new conceptual frameworks which are more in tune with the objectives of a truly sustainable agriculture.

The 'sustainability' concept has thus prompted much discussion about the need to make major adjustments in conventional industrial agriculture policy to become more environmentally, socially and economically viable and compatible. As a result, several possible solutions to the environmental problems created by industrial agriculture's emphasis on capital and technology-intensive farming have been proposed and research is currently in progress to evaluate alternative systems. The main focus lies on the reduction or elimination of agro-chemical inputs through changes in management to assure adequate plant nutrition and plant protection through organic nutrient sources and integrated pest management. But at the same time, as we have seen above, efforts are being made to introduce synthetic biology and to intensify genetic engineering that may result in opposite trends.

That is why the *Crossroads* report points out that, although hundreds of more environmentally-prone research projects and technological development attempts have taken place, and many lessons have been learned, the thrust is still highly technological, emphasising the suppression of limiting factors or the symptoms that obscure ill-producing agro-ecosystems. The prevalent 'scientific' explanation is that pests, nutrient deficiencies or other factors are the cause of low productivity, as opposed to the view that pests or nutrients only become limiting if conditions in the agro-ecosystems are not in balance. This leaves room for those proposing solutions such as 'synthetic biology' to emerge.

For this reason, there still prevails a narrow view that specific causes are the ones that affect productivity, and that the limiting factors must be overcome through the intensification of new technologies such as synthetic biology, genetic engineering and nanotechnology. This view has diverted agriculturists and 'practitioners' from looking at the entire agro-ecosystems and realising and appreciating the context and complexity of agro-ecological processes, thus underestimating the root causes of agricultural limitations. This is a distortion that agricology must overcome.

On the other hand, the science of agroecology, which is defined as the application of ecological concepts and principles to the design and management of sustainable agro-ecosystems, has provided a framework for assessing the complexity of agro-ecosystems. However, the epistemology underlying agroecology requires that we must go beyond the use of alternative scientific practises as such and develop new conceptual frameworks that incorporate IKS and promote agro-ecosystems with the minimal dependence on high agro-chemical and energy inputs. Such conceptual frameworks should encourage complex agricultural systems in which ecological interactions and synergisms between biological components take place and provide the mechanisms for the systems to generate their own soil fertility, productivity, and crop protection instead of artificially 'mimicking' them. These frameworks can still be found in the practises of the indigenous agriculture as practised in many communities and are a good basis on which an agricological system can be implanted on the ground to strengthen these organic practises in IKS and science. Therefore, there is a need to go beyond even agro-ecological principles to establish a wholly-integrated system of the new agriculture in the epistemology of Afrikology, as we shall see below.

We can thus look at the emerging principles of *agricology* as being concerned with the search to reinstate more ecological and natural rationale into agricultural production on the basis of the resilient practises that have been passed on from ancient times. This is because agricultural scientists and developers have disregarded this deep understanding of the nature of agro-ecosystems and the principles by which they function. Given this limitation, agroecology has emerged as the discipline that provides the basic ecological principles for how to study, design and manage agro-ecosystems that are both productive and natural resource conserving, and that are also culturally sensitive, socially just and economically viable [Altieri, 1995]. Agroecology therefore goes beyond a one-dimensional view of agro-ecosystems by taking into account their genetics, agronomy, edaphology, and so on, to embrace an understanding of ecological and social levels of co-evolution, structure and function. Instead of focusing on one particular component of the agro-ecosystem, agroecology emphasises the interrelatedness of all agro-ecosystem components and the complex dynamics of ecological processes [Vandermeer, 1995].

Agro-ecosystems are communities of plants and animals interacting with their physical and chemical environments. They have been modified by people to produce food, fibre, fuel, and other products for human consumption and processing. Agroecology is the holistic study of agro-ecosystems, including all environmental and human elements. It focuses on the form, dynamics and functions of their interrelationships and the processes in which they are involved. An area used for agricultural production, such as a field, is seen as a complex system in which ecological processes found under natural conditions also occur. These processes could be nutrient cycling, predator/prey interactions, competition, symbiosis and successional changes. Implicit in agro-ecological research is the idea that, by understanding these ecological relationships and processes, agro-ecosystems can be manipulated to improve production and to produce more sustainably, with fewer negative environmental or social impacts and fewer external inputs [Altieri, 1995].

The emerging 'science' and 'design' of agro-ecological systems of production, which is intended to mitigate the grave effects of industrial agriculture, is based on the application of the following ecological principles [Reinjntjes et al., 1992]:

- Enhancing the recycling of biomass and optimising nutrient availability and balancing nutrient flow;
- Securing favourable soil conditions for plant growth, particularly by managing organic matter and enhancing soil biotic activity;
- Minimising losses due to flows of solar radiation, air and water by way of microclimate management, water harvesting and soil management through increased soil cover;
- Species and genetic diversification of the agro-ecosystem in time and space; and
- Enhancing beneficial biological interactions and synergisms among agro-biodiversity components thus resulting in the promotion of key ecological processes and services.

These principles can, according to this school, be applied by way of various techniques and strategies, but this has the danger of straitjacketing this epistemology into 'methodologies', which can turn the existing 'sciences' into 'mono-disciplines'. Indeed, by referring to itself as a 'discipline', agroecology stands in danger of becoming an appendage of the current knowledge crisis and pushing us back into 'scientism', from which we are trying to escape, instead of enhancing dialogue and interaction between agro-ecological actors on the ground. If adopted, these different strategies and techniques will have different effects on productivity, stability and resiliency within the farm system, depending on the local opportunities, resource constraints and, in most cases, on the market. The ultimate goal of agro-ecological design should, in fact, integrate different agro components so that overall biological efficiency is improved; biodiversity is preserved; and the agro-ecosystem productivity and its self-sustaining capacity maintained. The goal is to design a quilt of agro-ecosystems within a landscape unit, not by just 'mimicking' the structure and function of natural ecosystems, but by enhancing their natural interconnectedness.

From a knowledge management perspective, the agro-ecological objective is seen by the new 'discipline' as providing balanced environments, sustained yields, biologically mediated soil fertility and natural pest regulation through 'the design' of diversified agro-ecosystems and the use of low-input technologies [Gleissman, 1998]. This is, in fact, turning agroecology into a rather technical system of management. This is because agro-ecologists misunderstand what is actually taking place in recognising that the intercropping, agro-forestry, permaculture and other diversification methods do 'mimic' natural ecological processes, when in fact the intercropping enhances natural processes just as nature enhances intercropping production in a 'dialectical' manner. The agro-ecologists also tend to 'technicalise' the process when they praise themselves that the sustainability of complex agro-ecosystems lies in the ecological models they follow, when in fact there is a natural process at work.

They believe that by designing farming systems that 'mimic' nature, optimal use can be made of sunlight, soil nutrients and rainfall [Pretty, 1994]. If this is the case, then we should not forget that current mainstream natural science disciplines, including agriculture and forestry as well as even 'synthetic biology', do 'mimic' nature in regarding themselves as 'scientific' vis-a-vis the social and 'human' sciences [Taylor, 1985]. It is true, therefore, to say that in designing systems or operations, human beings try to mimic nature in order to obtain the maximum efficiency in-built in it; but it is not true that we can mimic nature in order to improve it because nature has its own self-correcting and self-enhancing capacities, which human action cannot improve.

In our view, therefore, instead of artificially mimicking nature in order to improve and enhance natural processes, the agro-ecological management scientists must lead natural resource management towards the optimal recycling of nutrients and organic matter turnover, closed energy flows, water and soil conservation and balanced pest-natural enemy populations, which in the final process sets in motion the self-regenerative processes of nature. This strategy exploits the complementarities and synergisms that result from the various natural combinations of crops, trees and animals in spatial and temporal arrangements being put together [Altieri, 1994]. In fact, such combinations still exist in the theories and practises of indigenous knowledge systems, as we have seen. So what is required is the need for an interface between the agro-ecological scientists and the indigenous knowledge experts and practitioners by combining what is possible in their knowledge systems.

In essence, the agro-ecologists admit that the optimal behaviour of agro-ecosystems depends on the level of interactions between the various biotic and abiotic components. By assembling a functional biodiversity it is possible to initiate synergisms which subsidise agro-ecosystem processes by providing ecological services such as the activation of soil biology, the recycling of nutrients, the enhancement of beneficial arthropods and antagonists, and so on [Altieri, 1999]. Even though today there exists a diverse selection of practises and technologies, which vary in effectiveness as well as in strategic value such as those of a preventative nature and which act by reinforcing the 'immunity' of the agro-ecosystem through a series of mechanisms, it should not be forgotten that such technologies do exist in indigenous knowledge systems, although they have been undermined by the mainstream technologies. These indigenous technologies and strategies, if properly understood and applied, can help to restore agricultural diversity in time and space, including the use of intercropping, crop rotations, cover crops, crop-livestock integration, etc. Indeed, it is the indigenous ecological systems which claim the following features also recognised by the agroecology scientists:

Crop Rotations: This is a temporal and spatial diversity incorporated into cropping systems, providing crop nutrients and breaking the life cycles of several insect pests, diseases, and weed life cycles [Sumner, 1982].

Polycultures: This involves complex cropping systems in which two or more crop species are planted within sufficient spatial proximity to result in competition or complementation, thus enhancing yields.

Agro-forestry Systems: This is an agricultural system where trees are grown together with annual crops and/or animals, resulting in enhanced complementary relations between components increasing multiple use of the agro-ecosystem.

Cover Crops: These involve the use of pure or mixed strands of legumes or other annual plant species under fruit trees for the purpose of improving soil fertility, enhancing biological control of pests, and modifying the orchard microclimate [Finch and Sharp, 1976].

Animal integration in agro-ecosystems: This assists in achieving high biomass output and optimal recycling.

According to Altieri [1995], all of the above diversified forms of agro-ecosystems share in common the following features, which are also recognised by indigenous knowledge systems, which should not be subsumed under the agro-ecological science without acknowledgement of the indigenous knowledge.

- Maintaining vegetative cover as an effective soil and water conserving measure that is met through the use of no-till practises, mulch farming, and the use of cover crops and other appropriate methods;
- Providing a regular supply of organic matter through the addition of organic matter (manure, compost, and promotion of soil biotic activity);
- Enhancing nutrient recycling mechanisms through the use of livestock systems based on legumes, etc.; and
- Promoting pest regulation through the enhanced activity of biological control agents that are achieved by introducing and/or conserving natural enemies and antagonists.

These scientists also point out that research on diversified cropping systems does underscore the great importance of diversity in an agricultural setting. They also recognise diversity as being of value in agro-ecosystems for a variety of reasons [Altieri, 1994; Gliessman, 1998] among which are:

- As diversity increases, so do opportunities for coexistence and beneficial interactions between species that can enhance agro-ecosystem sustainability. This is a natural process;
- The prevalence of greater diversity often allows better resource-use efficiency in an agro-ecosystem. There is better system-level adaptation to habitat heterogeneity, leading to complementarity in crop species needs, diversification of niches, overlap of species niches, and partitioning of resources;

- Ecosystems in which plant species are intermingled possess an associated resistance to herbivores, as in diverse systems there is a greater abundance and diversity of natural enemies of pest insects keeping in check the populations of individual herbivore species;
- A diverse crop assemblage can create a diversity of microclimates within the cropping system that can be occupied by a range of noncrop organisms – including beneficial predators, parasites, pollinators, soil fauna and antagonists – that are of importance for the entire system;
- Diversity in the agricultural landscape can contribute to the conservation of biodiversity in surrounding natural ecosystems;
- Diversity in the soil performs a variety of ecological services such as nutrient recycling and detoxification of noxious chemicals and regulation of plant growth; and
- Diversity reduces risk for farmers, especially in marginal areas with more unpredictable environmental conditions. If one crop does not do well, income from others can compensate the loss or reduction.

However, the designing of agroecology and sustainable agro-ecosystems can be useful if the designing is interfaced with the designs of other knowledge co-producers. Most people involved in the promotion of sustainable agriculture aim at creating a form of agriculture that maintains productivity in the long term through 'designing' [Pretty, 1994; Vandermeer, 1995]. This can be achieved, according to these scientists by:

- optimising the use of locally available resources by combining the different components of the farm system, i.e. plants, animals, soil, water, climate and people, so that they complement each other and have the greatest possible synergetic effects;
- reducing the use of off-farm, external and non-renewable inputs with the greatest potential to damage the environment or harm the health of farmers and consumers; and a more targeted use of the remaining inputs used with a view to minimising variable costs;
- relying mainly on resources within the agro-ecosystem by replacing external inputs with nutrient cycling, better conservation, and an expanded use of local resources;
- improving the match between cropping patterns and the productive potential and environmental constraints of climate and landscape to

ensure long-term sustainability of current production levels;
* working to value and conserve biological diversity, both in the wild and in domesticated landscapes, and making optimal use of the biological and genetic potential of plant and animal species; and
* taking full advantage of local (indigenous) knowledge and practises, including innovative approaches not yet fully understood by scientists although widely adopted by farmers.

Thus, through the application of agro-ecological principles, the basic challenge for sustainable agriculture to make better use of internal resources can be easily achieved by minimising the external inputs such as chemicals used, and can contribute to the regeneration of internal resources more effectively through diversification strategies that enhance synergisms among key components of the agro-ecosystem. The ultimate goal of agro-ecological design is therefore to integrate components so that overall biological efficiency is improved, biodiversity is preserved, and the agro-ecosystem productivity and its self-regulating capacity are maintained. The goal is to design an agro-ecosystem that is intended not just to 'mimic' the structure and function of local natural ecosystems, but to enhance their regenerative power – that is, enhancing a system with high species diversity and a biologically active soil, one that promotes natural pest control, nutrient recycling and high soil cover to prevent resource losses.

In that way, we can conclude that agroecology can provide other forms of information and knowledge to farmers in a cooperative manner without such knowledge being a hindrance to the natural diversified agro-ecosystems that take advantage of the effects of the integration of plant and animal biodiversity. Such integration enhances complex interactions and synergisms and optimises ecosystem functions and processes, such as biotic regulation of harmful organisms, nutrient recycling, and biomass production and accumulation, thus allowing agro-ecosystems to sponsor their own functioning instead of them being 'synthesised' into artificial genetic processes.

The end result of agro-ecological design is to attain improved economic and ecological sustainability of the agro-ecosystem, with the proposed pest management systems specifically in tune with the local resource base and knowledge systems, as well as an operational framework of existing environmental and socioeconomic conditions. In an agro-ecological strategy, management components are directed to highlight the conservation and enhancement of local agricultural resources (germplasm, soil, beneficial fauna, plant biodiversity, etc.) by emphasising regenerative epistemology that encourages farmer participation, use of traditional knowledge, and adaptation of farm enterprises that fit local needs and socioeconomic and biophysical conditions. This need not be a 'methodology' but the encouragement of interactions, dialogues, and interfacing of different producers and actors with their different contributions into a holistic system of co-producers.

[E] Restoring Traditional Governance and Justice

The resilience of traditional knowledge systems and the ecosystems that sustain life are all interlinked to forms of restorative governance and justice. This is within the traditional holistic system of knowledge that has spiritual elements and which recognises the right of all people to participate in community life – beginning with the family. Thus, within the traditional knowledge systems are incorporated philosophies, theories, and practises of a political nature that are promoted by the communities through their political, social and cultural organisations. It is clear that no community can sustain itself as a human community unless they have created political and governance systems through which they can regulate their social and economic relations and ensure that justice is done to all within the community. Here the concepts 'governance' and 'justice' are interlinked because they do not accommodate antagonistic paradigms such as 'adversarial' or 'redistributive' systems. Such systems are holistic because they aim at a restorative justice at all levels and the need to restore balance on relations between humans, nature and the spiritual world of the dead and the unborn. This is especially so in Africa where traditional systems of governance have persisted for long periods of time and whose philosophies and practises have been passed on from generation to generation through the divine 'living word'. Even the colonial rulers found that in order to govern the Africans, they needed to work through their traditional institutions in the form of 'indirect rule' or neo-traditionalism.

Today, as the post-colonial states face collapse and many are referred to as 'failed states', there is a clear need for the traditional institutions of governance and justice to be restored and recognised in the same way their IKS are also recognised. As the modern states decline, there has been a new trend towards the revival of these institutions in which the people have trust [Nabudere, 2002]. But the resilience of these systems lies in their historical basis, which has enabled them to build up certain principles which sustain them on a popular basis.

In his study of African societies in his famous book, *The Destruction of African Civilisation*, Chancellor Williams summarised these different experiences of African political systems, which he gave the general name of the African Constitution. This constitution is a body of principles and practise, which he draws from the customary laws that governed the Black African societies from ancient times up to the time of invasions. He traces the lineage ties and corresponding responsibilities and age-set and age-grade systems as the earliest institutions through which the African Constitution functioned and out of which its democracy was born. According to Williams, these elementary elements continued right through 'tribal' societies and state forms such as kingdoms and empires [Ibid: 160-5]. He believes that a majority of African states operated on the principle of acceptance of common ancestry and the construction of the lineage system as 'the powerful factor' that provided the basis and incentive for the later formation of kingdoms and empires.

Out of these basic relationships that created the traditional political systems, Williams postulates some theories and principles of traditional constitutional law, and these spell out the fundamental rights of the African people. For instance, one of these principles is the position of the common people as the final source of power. The second is the recognition that the rights of the community of people are superior to those of individual members, including those of chiefs and kings. These principles can be found scattered in the different constitutions of the communities that did not accept chiefs and those that were in the kingdoms and empires as well [Ibid: 171-76].

It is through these principles, which are renegotiated from time to time, that African small farmers have been able to sustain their economies, including agriculture, on an ecological basis. Recent surveys and studies demonstrate that wherever traditional systems have re-emerged in the post-colonial period, they have shown more inclination to defend their rights to land and the products on those lands, including forests, rivers, lakes, and the cattle that feed on those lands. Pastoralism is one of the age-old economies of Africa and their persistence has demonstrated a deep linkage between the pastoralists and their cattle economies and the reciprocal relationship with the agriculturalists. Theirs is a spiritual system that links them to the animal world, where livestock is recognised as playing a spiritual-cultural role in their society.

A recent report compiled by the Overseas Development Institute (ODI) in the UK has revealed that mobile livestock, contrary to earlier theories, adapts to climatic change variability, allowing pastoralists to transform the seeming wastelands into productive assets. According to the report, strengthening pastoralists' land tenure and facilitating their representation as well as improving their access to education, enables them to address food security and by also producing for the market at low cost. The report notes that, at the same time, pastoralism does not negate diversification of rangeland use. It also does not negate other eco-friendly activities such as complex cropping systems, including intercropping, agro-forestry and rotational cropping, which are characteristic of extensive and shifting cultivation.

Thus, the common pool of resources is amenable to multiple enterprises, which should be recognised as contributing to human survival through collective action and management. These experiences, from organic farming by ordinary farmers, demonstrate that the world can feed itself without the constraints of the Hardinist and Malthusian false predictions that are based on unilinear thinking and epistemologies of a 'one-size-fits-all' ideology of 'mass starvation'. This is a disempowering ideology for communities, which deprives them and forces them to adopt strategies that cannot enable them to feed themselves without the altruism of the State and private capitalists acting as the 'donor community' who at the same time dominate the world's resources.

Therefore, there is a need to recognise the role of these traditional institutions in redefining democracy towards a more direct approach to self-administration instead of 'representational democracy'. Prof Ray of Canada, who has done a study of traditional institutions, has recommended a range of possibilities for the involvement of traditional leaders in local government. He has advocated that the relationship between traditional leaders and local, regional, and national government be interactive instead of being a top-down affair. According to him, in contemporary conditions, traditional leaders can legitimatise the State by acting on behalf of the State objectives of development and democratisation, while the State sets the terms of traditional leaders' legitimacy in the contemporary era and also by providing new frameworks and resources within which traditional leadership can operate [Ibid: 116]. This is the way to go if the modern post-colonial state is to avoid its rapid fall from favour by adapting to the changing conditions.

[F] FROM AGRICULTURE TO AGRICOLOGY

i. The 'science' of agroecology

The movement from industrial agriculture to a new agriculturally-sustainable system (agricology) integrates agroecology with other ecologically-friendly systems that have emerged in opposition to the mainstream 'scientific' agriculture as it is practised today. As we have seen above, the mainstream system relies on the input-output model based on the extensive use of agro-chemicals and mechanisation. Among a crop of emerging academics and social practitioners, agroecology is seen as a new science and discipline based on ecological principles of farming. The emerging system is less capital intensive and is based on the efficient use of locally available bio-resources. Instead of the input-output model of industrial agriculture, the emerging agro-ecological system relies more on cyclic movement of nutrients, water and other resources and in so doing promotes a high level of biodiversity in the system. It therefore offers a great deal of benefits in the form of sustainability, stability, security and higher cumulative productivity over time. As such, it is well-suited to small and poor farmers [Dharmitra, 2010: 10-11].

According to Altieri [1995], 'the science of agroecology ... defines as the application of ecological concepts and principles to the design and management of sustainable agro-ecosystems' as well as providing a framework 'to assess the complexity of agro-ecosystems'. Altieri continues that 'the idea of agroecology is to go beyond the use of alternative practises and to develop agro-ecosystems with the minimal dependence on high agro-chemical and energy inputs, emphasising complex agricultural systems to which ecological interactions and synergisms between biological components provide the mechanisms for the systems to sponsor their own soil fertility, productivity and crop protection'.

Due to the fact that the mainstream scientists have disregarded a key point in the development of a more self-sufficient and sustainable agriculture and its lack of deep understanding of the nature of agro-ecosystems, agroecology has become the discipline that provides 'the basic ecological principles for how to study, design and manage agro-ecosystems that are both productive and natural resource conserving, and that are culturally sensitive, socially just and economically viable'. Therefore, the discipline and science of agroecology 'goes beyond a one-dimensional view of agro-ecosystems – their genetics, agronomy, edaphology, and so on – to embrace an understanding of ecological and social levels of co-evolution structure and function'. Thus, while this school emphasises that agroecology is a 'holistic study' of the agro-ecosystems, including environmental and human elements, it still maintains that by understanding these ecological relationships and processes, 'agro-ecosystems *can be manipulated* to improve production and to produce more sustainably, with fewer negative environmental or social impacts and fewer external inputs' [Ibid.]. This is the basis on which the ecological principles are formulated, as we saw above.

But as we have already seen and remarked, the new discipline has the danger of falling into fragmentation since it claims to develop methodologies, techniques and tools for designing production systems which mimic nature. But what they try to mimic is not nature, for nature cannot be mimicked or copied: rather knowledge systems that already exist, which are in close relationship with nature and work well along with it to perpetuate the ecosystem in which nature and man thrive without damaging each other. This is because Altieri maintains that from a 'management' perspective, the agro-ecosystem objective is 'to provide a balanced environment, sustained yields, biologically mediated soil fertility and natural pest regulation through the design of diversified agro-ecosystems and the use of low-input technologies'. That is why the new agro-ecologists are now recognising that intercropping, agro-forestry and other diversification methods 'mimic natural ecological processes, and that the sustainability of complex agro-ecosystems lies in the ecological models they follow'. He continues: 'By designing farming systems that mimic nature, optimal use can be made of sunlight, soil nutrients and rainfall.'

This process sounds contrived by 'models' and 'methods'. Why should 'farming systems' mimic nature, when they are in fact part of nature, unless we are talking of the reductionist mainstream systems? This seems inevitable since it is only now that the new ecologists using their new 'science' are able to recognise that intercropping and agro-forestry and other diversification systems do in fact mimic nature. Therefore, the new ecologists must learn from indigenous knowledge systems if they are not to lead us astray through 'disciplines' that 'design' natural processes. That is why, later, Altieri wakes up to the fact that in order for 'sustainable agriculture' to maintain productivity 'in the long term' it must 'take full advantage of local knowledge and practises, including innovative approaches not yet fully understood by scientists although widely adopted by farmers'.

That means, although the new ecologists are mimicking indigenous knowledge systems without acknowledgement (and not nature, for nature cannot be mimicked unless one has some knowledge about it!), they still do not understand it. Moreover, the indigenous knowledge they are talking about is not 'local knowledge and practises'. IKS is universal knowledge and practises by all indigenous farming communities throughout the world since agriculture was discovered by Osiris and spread around the world. This misunderstanding is fundamental and must be overcome by the new ecologists if they are to begin to learn IKS and work with them rather that claim to be mimicking nature. Agro-ecologists cannot claim to 'provide guidelines to develop diversified agro-ecosystems' when these already exist in nature and are known by IKS. Furthermore, the new agro-ecologists cannot, like their mainstream scientists, claim to 'develop methodology that encourages farmer participation, the use of traditional knowledge and the adaptation of farm enterprises that fit local needs and socioeconomic and biophysical conditions', when they themselves are ignorant of IKS.

The indigenous knowledge systems have been in use throughout millennia and are in conformity with the way nature reproduces itself with agriculture as one such system. Therefore, if agroecology is to be truly a new form of agriculture, it must integrate itself and be validated by indigenous knowledge systems in order to be part of a holistic system of production and reproduction of food and other material needs on which humanity can survive.

In short, what agricology must do is to revive, rejuvenate, and restore organic relationships between the human beings; the plant world; the animal world; and nature, which sustains all of them. What we are involved in is not therefore a 'shift' in 'method', 'technique', or 'process'. What we are involved in is bringing about an epistemological revolution, but which goes beyond the existing academic disciplines of knowledge towards what currently is referred to as trans-disciplinarity towards a spiritually-based system of knowledge and wisdom that draws from our heritages of how agriculture was understood and what its role was supposed to be. This means also going beyond agroecology as a 'discipline' and 'science'. This revolution will move us away from the current economic systems based on capitalism to new economic systems that are based on a circular spiritual relationship that enables us to balance the economic needs of humanity and the animal world with those of nature on which we depend.

We have already demonstrated how modern capitalist industrial agriculture adopted systems of 'science', which began to undermine this balance. These systems were 'imitative' of nature and reductionist in their conceptions. However, dictated by individual desire to 'accumulate wealth' in the form of commodities as represented by the money-commodity, capitalism has gone wild and begun to destroy the agro-ecosystem on which we depend. This system of production, based on individual greed, has exhausted itself although it seeks to contrive 'new sciences' such as synthetic biology and genetic engineering to put in place a 'new industrial revolution' based on a global private expropriation of biomass for new forms of individual profit and accumulation. This attempt, as we have argued above, leads us to a doomsday scenario in which we shall all perish along with nature. But nature will extract revenge and will survive while humanity will vanish.

The indigenous knowledge systems about agriculture can help us to restore agriculture (as agricology) and to reverse the damage that modern industrial agriculture has done. As we have also noted, this knowledge spreading back to antiquity has not been fully destroyed by modern reductionist 'science', since it continues to exist and operate in practise and through innovation as we have noted above. These knowledge systems constitute the basis of some of the agro-ecological practises that are being embraced even by modern scientists. But IKS are different from the agro-ecological system, which is currently being practised in different parts of the world. These agro-ecological systems are all based on 'designed' and judicious use of available biological knowledge resources which follow ecological principles for the efficient management of nutrients, pests and water.

These systems offer the production of multiple crops at sustainable levels and at lower cost compared to industrial agriculture. However, there is room for further improvement based on close examination of the biological processes of soils and an understanding of the interactions among the various components of biodiversity systems, which IKS alone offer. What is important to note is that all these varieties and practises of agriculture are nevertheless based on traditional (or indigenous) systems of knowledge. As the editors of the *Third World Resurgence Journal* [2010] have observed:

'While it must never cease to tap the abundant and astonishing varieties of traditional knowledge held by local communities, it is not a closed system of knowledge. It is important to remind ourselves that there is no social group more innovative than small farmers and that the notion of 'traditional' being synonymous with 'unchanging' is a myth. Traditional farmers have, on the basis of their experience, been more open to absorbing new knowledge and practises from diverse sources than is commonly supposed [Ibid: Editor's Note].'

This means that the inherited traditional knowledge throughout the world is more open to scientific knowledge if that knowledge is grounded in the reality of a self-sustaining nature and environment. This is why recent scientific attention has focused on endogenous biological processes within soil systems, which offer opportunities to increase agricultural output with minimal dependence on external inputs. These practises can be accepted by small farmers because they confirm what they have been doing. According to Dharmitra [2010]:

'There is growing evidence to show that a variety of crops can be cultivated more abundantly and at lower production cost by managing and intensifying endogenous soil systems using existing genetic potential of crops, achieving a substantial increase in output from 50-100 per cent. By comparing average yields of organic versus conventional or low-intensive food production and estimating the average yield rations (organic; non-organic) for the developed and developing world, Badgley and others [2007] modelled a global food supply which could be grown organically on the current land use [Ibid.].'

Dharmitra continues to observe that the model estimates produced by these authors indicate that organic methods could produce enough food on a global per capita basis to sustain the current human population, and potentially an even larger population, without increasing the agricultural land base. They also calculated the amount of nitrogen potential available from fixation by leguminous cover crops used as fertiliser and based on the data collected from temperate and tropical agro-ecosystems, concluding that leguminous cover crop could fix enough nitrogen to replace the amount of synthetic fertiliser currently in use [Ibid.]. But what Badgley and his colleagues have produced merely demonstrates what the small farmers are able to do even beyond such expectations and predictions. Evidence from India – where the Green Revolution had devastated farmlands and pushed small farmers to marginal and poor lands – has shown that lands which had been degraded by chemicals from industrial agriculture could be reclaimed and put to productive use.

In confirmation of the above scientific discoveries about the productivity of indigenous production, the United Nations studies by UNEP-UNCTAD on 114 cases in Africa revealed that a shift towards organic agriculture produced increased yields by 116 per cent. The study pointed out: 'moreover, the positive impact endures as it is based on strengthening the five types of capital in farming communities: human, social, natural, financial, and physical'. Another study released by the UN Special Rapporteur on the Right to Food revealed that small-scale farmers could double their food production within 10 years in critical regions of the world by using ecological methods. Relying on these findings, the Rapporteur called for a 'fundamental shift towards agroecology as a way to boost food production and improve the situation of the poorest [UNCTAD, 2011].

ii. Organic agriculture vs. Green Revolution

This was revealed at a conference on organic farming held at the University of Agricultural Science in September 2009 at Bangalore on the theme 'Outstanding Organic Agriculture Techniques'. Papers and PowerPoint presentations presented at this conference by poor farmers in their own mother tongues produced evidence to this effect: that degraded lands could indeed be reclaimed through proper organic farming. We shall give this evidence from India to demonstrate how widespread indigenous knowledge has been used throughout the world to maintain the soil and agriculture throughout history and how quickly natural processes can be restored if the correct knowledge is used to reverse the effects of modern 'scientific' agriculture.

A farmer from the Tamil Nadu Organic Farmers' Movement presented evidence which demonstrated that organic farming techniques could rehabilitate and revive lands that had been affected by salt water after the tsunami hurricane disaster of 2004. She pointed out how her association had tried out these methods by first ploughing out the salt-affected lands thoroughly as they created trenches along the fields. These trenches were packed with coarse materials to absorb saline water and to encourage an increase in the microbial population. On the ploughed lands daicha seed was broadcast to provide the required biomass and to initiate soil activity. Farmers prepared *panchagavya* (a popular recipe for dramatically increasing the population of beneficial microorganisms). Large numbers of vermicomposting units were also set up to produce additional organic manure to save on input costs. According to her:

> 'Through these methods, in less than a year, fields were replanted with paddy successfully to the absolute amazement of government officials and international agencies who had earlier thought of solving the problem by simply dumping huge quantities of gypsum in them [Alvares, 2009: 18].'

In another presentation, a small organic farmer made a presentation on how one could grow food on severely eroded lands. She demonstrated how she and another organic farmer had taken up for rehabilitation such a land with the aim of turning it into a natural forest and wilderness. When they first took up the land it had no top soil and was a compact mass of only pebbles and laterite. The main challenge they faced was how to build soil, as cultivated plants need at least six inches of good soil to grow. The ecological rules of their association forbade them from bringing in soil from outside or even purchasing compost or manure. All biomass had to be grown at the site and all resources had to come only from the homes and surroundings. Moreover, in their case no worker was employed.

Their process of building up the soil included immediate measures to protect the land from further assault by the elements; creating water bodies and contour bunds; establishing pioneer vegetation to produce biomass at site; creating raised beds and then building up and maintaining organic matter by creative use of local resources. She demonstrated through slides how this was done and how this could be replicated anywhere. Please notice that the activity here does not try to mimic nature *but replicates and rejuvenates* the natural resources within the environs of the damaged lands. This replication was possible so long as the principles of ecological restoration of soil were strictly followed. Once the soil was restored, the next stage of working with the land was to ensure that no more soil was turned.

Another farmer from Pradesh state made a presentation of how, following the theories of one, Professor Dabholkar, a popular scientist from Maharashatra state who worked with organic farmers to produce miracle crops without using chemicals and poisons. The farmer was able to demonstrate how he was able to cultivate 150 varieties of crops which provided food for a family of five. He showed how they were able to produce water elixir and soil elixir that met all the problems faced with current-day soils which had been ruined and rendered sterile by chemical farming.

A highly-respected and knowledgeable organic farmer from Tamil Nadu explained how the problems faced by all famers, whether organic or conventional, were the prevalence of troublesome insects, weeds and plant diseases. He demonstrated the techniques which farmers can use to solve most of their insect and plant disease problems. Having previously worked with chemicals on his farm, he was able to demonstrate from his experience as an organic farmer how to create microorganism-enriched mixtures (which he called MEM) that control soil-borne diseases, nematodes and root grubs. He had made large numbers of recipes from leaf extracts, buttermilk, waste fish and egg extracts, panchagavya and some unique other solutions which he called *fruit gaudi* and *archae*. Through use and multiplication, these recipes and techniques are today known and freely available to other organic farmers.

Presentations by other farmers dealt with diverse issues and problems concerning organic farming, which they shared with the other farmers. These included solutions and responses to problems encountered by all farmers such as how to recognise plants' own natural defence mechanisms and how it is best to work in tandem with the plants' own defence strategies than to attempt to bypass them with deadly chemicals. Unless these defence strategies are understood, a lot of mistakes will be made and the farmers believed that it was due to such ignorance of these defences that there was a resort to chemicals. They warned that continuing to use these deadly sprays and chemical poisons would suppress these natural defence mechanisms and damage the products produced through these means since the pesticide residues were bound to remain after harvest as most pesticides were non-biodegradable.

There were presentations which demonstrated how weeds can be an index of the quality of the soils and how conventional agriculture obsessed with being clean has been engaged in a permanent warfare against weeds, which leaves the soil exposed and thereby encourages the loss of moisture and soil erosions. The farmers called this a 'do-everything-culture', where the farmer is constantly doing unnecessary things to keep the land 'clean' and eventually losing out. The farmers called weeds 'misunderstood plants' which are constantly fought with herbicides and weedicides. They called such destruction nonsensical since by the selective introduction of certain species of green manuring plants a full control over weeds can be gained. Termites were also shown to be of importance to farmers because they help the soil to function better because they are known to eat any form of lingo-cellulose (paper, wood, jute, cotton) into pieces. Furthermore, the bellies of termites have a goldmine of microbes that are rich in sources of enzymes for converting lignocelluloses into organic manure and biofuels naturally. It is also by these means that termites assist forests to survive by helping in recycling dead materials.

Some farmers also demonstrated how to use herbals as repellents of unnecessary insects. Hence, organic farmers are prohibited from using chemical sprays that kill insects. For them killing insects is prohibited; only repelling is allowed. Organic farmers do not, in fact, refer to such insects as 'pests'. In nature, only insects and not pests are to be found. They are called 'pests' to justify their being killed! The farmers have also a guideline for locating herbal repellents – those plants which are not eaten by goats or cows; plants that yield milky saps; plants that have a bitter taste such as neem trees; plants that give off a bad odour; and plants that are poisonous [Ibid.].

These experiences from organic farmers demonstrated that small famers using organic farming methods can feed the world. According to Claude Alvares, 'Organic agriculture holds the key to long-term, successful, self-sustaining agriculture. Organic feeding can feed the world [Alvares, 2009: 8-10].' This is what he regards as the central message that came out of the pioneering, self-confident, organic farmers in Asia. At the conference referred to above, while the small 'uneducated' farmers were able to confidently present the evidence of their engagement in organic farming, they ensured that academics from universities were 'kept at bay', since many of them did not have any contribution to make on the subject while the organic farmers took to the centre stage: 'It was an exhilarating experience to see so many ordinary men and women, almost all practising organic farming, many without the so-called benefits of even a college education or English, delivering with aplomb PowerPoint presentations based on their own field experiences [Ibid: 8].' Agricultural scientists from universities who attended some of the sessions, according to Alvares, 'had to concede that in many areas-like the use of beneficial microbes to create living soils, preparing economic seed materials, restoring degraded farmland –the organic farmers were far ahead of the academic community in using innovative methods to solve problems considered extremely serious (or even hopeless) by conventional farmers' [Ibid.].

What the small farmers were trying to do was to restore the more than 40 centuries of organic farming, which modern industrial agriculture had destroyed within a few years of the so-called 'agricultures of permanence', which had unleashed a Green Revolution of hybrid seeds, mechanisation and chemicals on the world. This introduced enormous instability in all aspects of farming from seed availability to harvest disposal, which, as we have seen, led to farmer suicides in India. The Green Revolution not only began to kill the soil and the seeds, it began to kill the people as well, not only due to starvation amidst plenty but also by pushing the indebted small farmers to hang themselves. This form of conventional agriculture was leading the world to the 'self-destruction' of humanity and the universe itself.

The power of the rejuvenation of nature, given the right conditions, which the farmers out of centuries of wisdom inherited and accumulated by human experience and continued advocacy of the 'word' from generation to generation, has enabled parts of the adverse consequences to be reversed. The small farmers, out of their wisdom passed on from generation to generation through the 'living word', have been able to 'do wonders' and managed to enhance soil quality by reverting to traditional mixed farming using as many as 17 different types of crops from cereals and millets to pulses and oil seeds on one acre of land [Ibid.]. They have, by disproving the 'unnatural assumptions of modern industrial agriculture, been able to demonstrate through knowledge and action that, in nature, plants left on their own are able to "secure food for free"'.

They were able to prove this by pointing out that forests everywhere in the world are able to 'feed themselves' and that 'no human being is able to create a natural forest'. Forests manage all their nutrients themselves and are able to generate and store them without human assistance and can, as a result, survive and prosper on a millennial basis. But the urbanised classes who are alienated from nature have, under modern 'scientific' knowledge systems, come to wrongly believe that plants, which are part of nature, cannot grow or prosper without the assistance of chemical fertilisers supplied by huge agribusiness corporations from which they make huge monetary profits. Such contrived, reductionist approaches cannot be scientific, nor can they sustain human, animal, plant life and the natural ecosystem in general. They are destructive and must be stopped and replaced with self-sustaining systems that draw from human knowledge that learns from nature and sustains it along with other beings on the planet and the ecosystem.

The Popular Knowledge Women's Initiative (PKWI) of eastern Uganda, through their own experience based on the Aduso philosophy and indigenous knowledge systems, argue that the principles on which organic farming is based, are enshrined in the epistemology of Afrikology and agroecology (see below). They point out that these principles are embodied with the spiritual ethos of Mother Nature. The practise of what they call 'enterprise mix' (see below) enables them to engage in agricultural cultivation alongside animal husbandry. This 'mix' improves rather than works against nature. Under this practise, each farmer manages several co-related enterprises, each enterprise being separate but depending on the other. Through this interlinkage, Mother Nature remains real, untouched but reinforced, and enhanced by human activity.

As we noted in the Preface, as part of the effort to rediscover, unlock, rejuvenate mainstream – and validate indigenous – knowledge systems practised and promoted by Aduso, PKWI has introduced the model of Key Farmers' Trainers (KFTs) who, themselves, are leaders and practicing farmers. This development arises out of the community's continued policy of engaging government research organisations such as the National Agricultural Research Organisation (NARO) and the National Semi-Arid Resources Research Insitute (NaSSARI) in helping to validate their ideas and practises drawn from their traditional knowledge systems to good effect. Apart from that, the KFTs receive research outputs from NARO-NaSSARI at the government research centre at Serere and are able to read and internalise them and then develop message scripts, which are taped or recorded and then passed over by KFTs to the farmers. The trainers alert 'farmer listening rings' to listen to the messages, ask questions, and then reinforce what they have heard by putting the ideas into practise. In so doing, they are able to actualise the ideas through *doing, using, and interacting as a mode of incremental learning and innovation.* A formal partnership between the PKWI Community Initiative and NARO, particularly NaSSARI, has been agreed to allow for the joint development and sharing of knowledge in the future.

Through this approach, PKWI community initiatives have used their indigenous knowledge to instil in its members the spiritual instinct of diversity since the practises are looked at by the communities as part and parcel of the diversity of the environment. Consequently, the impact which their activities have on their own well-being has also a positive impact on the environment (and Mother Nature). This spiritual connection compels each member to have a better social environment that is compatible with the wider agro-ecosystems in which nature is not disturbed. The concept of unity in diversity is what has kept the PKWI community project going in unity, despite the fact that it is engaged in diverse economic and social activities able to contribute to a healthy environment, which is necessary for a new approach to a holistic transformation.

iii. A Transformative Energy Systems

A new approach to agricultural production in the form of agricology requires a new policy and practise on a self-renewing energy system. This is because we are trying to get away from the use of a single energy system based on fossil oil that is behind the economic and ecological crisis we are facing today in the global economy. Therefore, we need to adopt a transformative energy system that can enable us to overcome the crisis as well as restore our balanced relationship with nature. Such a new system requires the adoption of an entirely new end-use economics that Amory Lovins advocated in the 1970s, but which now requires immediate implementation [Lovins, 1977].

The 'soft path' Lovins advocated involved seven elements: *First*, adopting a flexible, diverse mix of energy use, which is specific to local conditions such as solar, wind, biomass, hydro, as well as limited fossil fuel use. This means that we must make an attempt to dissolve the current monoculture economies in favour of flexible, diverse mixed-energy uses as well as turning attention from energy supply towards energy use. *Second*, placing primacy on renewable energy sources, which depend on current diverse sources, instead of using stored up 'natural capital' in fossil fuels exploitation. *Third*, converting to efficient *end-use* and encouraging conservation of energy instead of wasteful consumption. *Fourth*, saving energy that is matched to the task at hand in both quality and scale by constructing and designing structures that save energy. *Fifth*, constructing participation-oriented structures in both production and consumption by having recourse to diverse mixes that involve decentralised production and peoples' active involvement in conserving and flowing with natural processes. *Sixth and finally*, adopting a people-intensive development and job-creation approach. Unlike the old industrial capital-intensive production and consumption patterns, soft energy paths of renewable energy sources and conservation promote labour-intensive approaches that create jobs. These soft paths, in effect, promote quality instead of quantity production aimed at *dematerialisation* by 'doing more' with 'less' and by substituting information for materials and energy. They also promote decarbonisation by moving to cleaner, less pulling sources of energy [Milani, op. cit. 115-19].

While it is true that the new soft energy paths must draw from the experiences of the old industrial system, the new paths must also draw from the positive holistic experiences of what Ivan von Sertima had called 'the lost sciences' of pre-capitalist knowledge systems [von Sertima, op. cit.]. The principle behind these 'lost sciences' is the need to structure our economic activities around the *circularity* of the universe and ecosystems. We must avoid adopting systems because they are technically 'efficient' and forget the efficiency that is built within the circular processes of nature. Thus, when we examine the soft paths we must look for those circular natural processes that exploit the forces of nature without harming the sources of those forces. This applies to the diverse mix of energy use, which is specific to local conditions such as solar, wind, biomass, and hydro as we have noted above.

For example, the application of solar energy must begin with what Milani has called *passive solar*. This is the simplest and most decentralised form of solar energy. This involves the simple capturing of the natural warmth of the sun by intelligently designing and situating buildings in such a manner that they capture *more* of the circulating warmth with less cost to the consumer. In that way, the end-use of the energy is deeply embedded into, and widely distributed over, the built environment without obstructing its flow. Thus, passive solar can provide not only heating but also cooling capacity and power for the natural ventilation [Milani, op. cit. 122]. *Wind power* is a circular renewable energy that has proved less costly even under current production conditions in countries such as Denmark. The current nuclear energy crisis since the Fukuyama nuclear meltdown has forced countries such as Germany to consider recourse to wind energy as a safer energy system. Countries such as these with a lot of economic resources should go deeply into solar energy end-use as quickly as possible. The Danish experience demonstrates that people themselves can influence policy and bring about changes in the energy sector in the right directions.

Other forms of soft paths to new energy strategies are small plants of *hydro, biomass/biogas and geothermal energy*, which can contribute greatly to community self-reliance, as well as hydrogen, which can be used as a fuel or means of storage. These can be small-scale and decentralised because the equipment to produce them is as economical on a small-scale as on a large one. Hydrogen has circular potentialities because it can be produced from water by using rooftop solar collectors as a power source and can be stored for use on a community – or even household – basis. *Flywheel batteries*, which store energy mechanically rather than chemically, can provide greater autonomy for home- and community-based electrical storage. Experiments in Ireland have demonstrated that small combined electrical generation systems consisting of a wind generator and a bank of PV cells coupled to an engine running on biomass can be developed.

The new decentralised energy economy will demand new kinds of load management due to the fact that the consumers are also the producers of energy. This is because the grid serves the function of a battery for storage for the small producers. The adaptation to daily weather and seasonal climate conditions becomes important if combined with the information and communication activities so that the same cable that provides buildings with computer and telecommunications services can also provide energy management information and services, while also providing the municipal utility with a means of coordinating supply and demand. This assumes that the telecommunication system is as community-oriented and community-controlled as the energy system, which is now becoming possible globally, and that the telecommunication system would augment rather than undermine face-to-face communication [Ibid: 123-24].

These changes would imply a radical change in economic management and lifestyles to *qualitative end-use* and energy-service approaches to energy. This demands an ethical commitment to the way we address our real needs instead of a conspicuous consumption characteristic of the now collapsing non-ethical and self-interested economy. This requires the individual questioning of the *quality* of the human needs and consideration for other needs, which rely on the same resources, always recognising that not all needs are equal. The ethical considerations will take into account considerations such as 'do I really need it?' 'Are there other people who might need it?' 'Are all these needs healthy and transformative or regenerative of nature on which we are depending?' This means we must constantly distinguish between *wants* and *needs* and the social-ethical implications of those choices. The new path calls for a self-administered and self-regulated community economy.

iv. The seed as money

A critique of the modern monetary system, especially in its capitalist financial phase, is that it relies on commodity-money to act as a 'store of value' for private accumulation, and as an end-in-itself [Milani, op. cit. P. 8]. This is why the money commodities such as gold and silver came to dominate in the European late mercantile era and the beginnings of modern capitalism. Under these systems, money became an end in life to be desired and fought for by all means called 'competition'. But as Zarlenga and others have observed [Zarlenga, 2002; Milani, 2000], money in its origin in Egypt was looked upon as a social means of exchange and payment rather than an economic commodity, which came to characterise Asian and later European *abstract* forms of money. In Egypt, anthropologists have observed, money had, according to Zarlenga, acquired its function more by its use in social ceremony than as barter. Here standardised payments for brides and blood money for injuries and deaths were its main uses.

The use of cattle as an early monetary standard, according to Simmel, lay in its uniqueness of extending and diversifying human interdependence while excluding everything that was personal and specific. Money in this form and function distanced 'self' from 'other' and 'self' from 'object', thereby 'generating within the individual the dissidence of self-sufficiency and alienation'. The individual under these circumstances relates to the social whole as one social power confronting another 'since (the individual) is free to take business relations and cooperation wherever he likes'. In this form, money drives a wedge between 'possessing' and 'being': 'Through money, man is no longer enslaved in things'. Thus, the elimination of the personal element of exchange, according to Simmel, 'is the gateway to human freedom'.

This is why the Nuer of Southern Sudan drew a marked distinction between 'the money of work' and the money acquired through the sale of cattle or 'the money of cattle'. This dichotomy was balanced by another distinction between two sorts of cattle: purchased cattle or the 'cattle of money' and cattle received as bride price or 'the cattle of girls'. Together these four categories with several subsidiaries played a prominent role in determining relations of autonomy and dependence among the Nuer. In a somewhat similar vein, in his study of the Pokot moral economy and their attachment to cattle in what he called the 'Cattle Complex', the anthropologist, Herskovits, demonstrated that this attachment was shown in the identification and affection with the animals and a dislike with killing them except for rituals. This is because, according to him, cattle were associated with birth, death, and marriage ceremonies, with special customs and taboos relating to their milk. Because of this complexity, cattle had become a dominant element in their culture, and by extension, to the cultures of other pastoral communities in East Africa. Thus, cattle in this understanding were regarded as a seed that perpetuated the lives of the people in their social and cultural relations with cattle. As the cattle reproduced themselves so did the human beings in a circular natural way. In the new circular economy of agricology, money should be removed to its centrality as an exchange commodity to become a social relation, through which people can establish social, cultural and economic relations by sharing and/or giving gifts in the form of goods and services to one another to meet their needs.

[G] Afrikology and Agricology

'Agricology' is a new concept for a renewed agriculture and is a reminder of the need to adopt new forms of knowledge based on new epistemologies that go beyond existing mainstream Cartesian epistemologies on which the current 'disciplines' and 'scientific methods' are based. This new epistemology we have called 'Afrikology' [Nabudere, 2011]. These two synthetic concepts recognise that while we must pursue agro-ecological organic farming and other forms of restorative agriculture on a new basis, we must also recognise that agroecology must be transgressed by the new epistemology that grounds the economy in continuous and holistic knowledge systems that are rooted in the knowledge and practises of living communities, which they have inherited from ancient times and wisdom. We have pointed out that the abbreviation *Afri-* in Afrikology is not an ethnic labelling but the recognition of the *locus* from where the epistemology and knowledge system originated in the *Cradle of Humanity*, which happens to have been situated in Africa. We also used the concept logos, not because we did not have an African word to express the same thing, but because it is comes from Greek, which is organically linked to the linguistic expression from the same Cradle [Diop, 1980; Bernal, 1984]. *Afrikology as an epistemology is therefore a universal system*, as we shall see below.

Therefore we argue that agroecology in all its forms must be combined with the holistic epistemology of Afrikology that is linked to the IKS that informs most of the human experiences and concepts from the origins of humanity and how they conceived the universe. Hence, the concept, agricology, is drawn from a synthesis of agroecology and Afrikology since the two are combined in *theory* and *practise*. We have already seen that agroecology in the form of organic farming is based on the strength of the small farmers who, through replication and rejuvenation, try to restore nature by the act of the wisdom incorporated in indigenous knowledge systems that has been perpetuated from generation to generation through the 'living word' of the ancestors through languages. This ancient worldview believed in a circular reproduction of the seed in human beings and in the plants, which came to be expressed in spiritual terms as a trinity of birth, death, and resurrection.

The new holistic epistemology and knowledge system must be spiritually and morally-based in a system of an ever self-recreated universe order to guide our actions, which must be in conformity with nature. The spiritual revolution that we need is one that enables us to recapture the African philosophy of *Ubuntu*, which calls for complementarity and the recognition of the 'Other' as an organic part of the 'Self' as expressed in the dictum: *I exist because you exist*. This philosophy of mutual recognition has its origin in the ancient-African cosmic principle of *Ma'at* (or 'connective justice').

According to Jan Assmann in his book, *The Mind of Egypt* [1996], '*Ma'at* (or connective justice) is the principle that coalesces individuals into communities and at the same time gives their actions meaning and direction by ensuring that the good is rewarded while the evil is punished. It is a *moral principle* that insists on the need for *continuity and memory* (or recollection) because the ancient Africans (Egyptians) believed that *memory* and mutually supportive *action* depended on each other because both provided conditions for peace and stability. Connective justice in this sense prevailed in a universe where justice held individuals together by morally connecting consequences with deeds. This is what made justice "connective". *Ma'at* (or *Ubuntu*) also knits "the world itself into a meaningful world". Therefore, when connective justice stops functioning and evil goes unpunished, then good no longer prospers and the world becomes "out of joint" [Ibid: 132-3].'

This principle is also expressed in the Acholi concept of *piny rac*. According to the Acholi philosophic discourse, a situation becomes *piny rac*, in the words of Okot p'Bitek, the Acholi poet and philosopher, when 'the whole thing is out of hand', and 'when the entire apparatus of the culture cannot cope with the menace anymore'. This Acholi concept is based on the belief that society is in harmony when all elements of existence in the culture are in balanced relationships. If the harmonic balance is broken, then everything becomes *piny rac* – and 'out of joint', which requires a rebalancing by society carrying out rituals that promote reconciliation [Okot, 1971: 62].

From these philosophies and principles we can draw some useful lessons that can enable humanity to consolidate the positive gains that they have made from their recent fragmented existence by putting them in balance with the ancient, but still existing, wisdom such as *Ma'at* or *Ubuntu*. We can also link this into a dialogue that emanates from the discourse arising from the 'crisis of reason' of the modern sciences and refer to the consequences of that crisis to create a better moral world with the support of ancient wisdom. We have the memory which can enable this dialogue to take place so that we can in the spirit of *Ma'at* 'connect' *memory* to mutually supportive actions into harmony: if we can do that successfully, we shall have embarked on the path of a new civilisation which can *'rehumanise man in society'* [Diop, 1980].

The African communities in the Iteso region of Uganda represented by the PKWI community initiative believe that the restoration and rejuvenation of the indigenous knowledge offers the only hope for the survival of humanity. They also assert that it is only after the recovery and restoration of IKS that there can be an interface with the current knowledge systems, including modern science, leading to a reconvergence with nature. *They regard the recovery of IKS as vital to humanity's survival since it helps the transcendence of the living word from generation to generation in the communal beliefs of a circular reproduction of nature, life and the agro-ecological seed through the concept of life, death, and regeneration (resurrection).* The community derives its philosophy of rejuvenation and recovery of IKS by recognising the role played by their great and famous ancestral spiritual domain or custodian, 'Aduso'. They recognise that it is from the spiritual strength of 'Aduso' that the PKWI community got their inspiration to reorganise the community.

According to the community's orally-perpetuated memory, it was way back in 1946 when the British colonialists attempted to dismantle the Aduso indigenous philosophy. This philosophy, based on Iteso indigenous knowledge, advocated the continuation of indigenous agriculture by the cultivation and use of the root crops and plants called *Ikorom*, of which the British colonialist 'modern agricultural science' disapproved, preferring the planting of cassava as a 'modern' crop. Aduso, who was the source of inspiration of the Ateker Clan, as early as that time, challenged the colonialist introduction of cassava and campaigned for the continued cultivation of *Ikorom* and *Edusa* crops, and plants. The colonialists hit back by destroying the Aduso shrines in order to promote their new knowledge of agriculture. They also introduced songs demonising Aduso, which were taught in all colonial schools in the region and neighbouring districts. This song has been reproduced in the Preface to this monograph.

Although Aduso and her campaign were demonised by the colonisers, the spirit of Aduso never died because it is from this spirit that the PKWI community, through the memory of the resistance, were able to recover and resurrect the indigenous knowledge and philosophy of Aduso. This was proof of the validity of Afrikology because it is through the connectivity and continuity of the IKS through the 'living word' (as conceived in the Heart of humanity) that the community was able to mount a recovery and restorative process of the original word and philosophy preserved from generation to generation by the Iteso people. According to Reverend Sam Ebukalin, the promoter of the PKWI community initiative:

'By combining the nature of both the spiritual and objective knowledge systems of IKS, the Aduso spirit that could never be eroded, not even destroyed, is now being actualised in PKWI as a development model expressed in the concept: *Re-entering the Rural Agenda*. This model sets on the ground a natural ancestral option to re-build the local economy through the 'enterprise mix' of cultivation and animal husbandry – each depending on the other. Through this model, the Aduso spirit has once again found a mouthpiece through the actualisation of its philosophies and inspiration. The social environment can now be a reality to the communities that PKWI operates with and the partners that wish to associate with PKWI [*Notes for inclusion in the monograph* – dated 23-08-2011].'

As we have seen above and in the Preface, this practical philosophy based on indigenous knowledge systems of the African people, has informed PKWI's theories and practises in agriculture and animal husbandry, which lend easily to the new agricology based on the epistemology of Afrikology. This community approach confirms that it is through the Oral Word that continuous knowledge is passed on from generation to generation and used for sustenance of the language communities. Scientific knowledge is a tentative discovery, which is valid for the purpose for which it was researched, so long as it does not conflict with nature. Such knowledge can be added to, and become part of, indigenous knowledge to become part of the permanent store of human knowledge passed on to future generations. As the Iteso case also demonstrated, a science of astronomy is part and parcel of indigenous knowledge that goes back to the year 300 BC (*see Preface*).

We therefore, do not agree with those who advocate *trans-humanism*, which promotes the idea of new technologies enhancing human qualities of understanding, because humanity possesses sufficient technologies in their cultures which enhance our self understanding and the workings of nature without seeking to destroy them with reductionist techniques intended to control it and create new 'human beings'. Indeed, we can already discern the rejection of these anti-humanist viewpoints from a new, integrated, holistic science, which the quantum revolution has already drawn attention to. These new understandings have created conditions for the reconvergence of man with nature. According to the humanistic scientist, Ruth Nanda Anshen:

> 'Science now begins to focus on the convergence of man and nature, on the framework which makes us, as living beings, dependent parts of nature and simultaneously makes nature the object of our thoughts and actions. Scientists can no longer confront the universe as objective observers. Science recognises the participation of man with the universe. Speaking quantitatively, the universe is largely indifferent to what happens to man. Speaking qualitatively, nothing happens in man that does not have a bearing on the elements that constitute the universe. This gives cosmic significance to man [Anshen in preface to Chomsky, 1986: xi-xii].'

Anshen adds that the current 'scientific method' and Cartesian epistemology that looks at living organisms as mechanical objects to be manipulated, no longer meets these needs. The convergence that Anshen refers to requires the use of both the *intuitive* and the *empirical*, since such a *combination* of knowledge and wisdom gives us meaning in life in addition to *information*. Accuracy, which the 'scientific method' seeks, is not the same as truth. Therefore, method and object can no longer be separated. A common bond links man, animal, plant, and galaxy – 'in the unitary principle of all reality; for the self without the universe is empty' [Ibid.]. She adds:

> 'This universe, of which we human beings are particles, may be defined as a living universe, its respiration being only one of the many rhythms of its life. It is evolution itself. Although what we observe may seem to be a community of separate, independent units, in actuality these units are made of subunits, each with a life of its own, and the subunits constitute smaller living entities. At no level in the hierarchy of nature is independence a reality. For that which lives and constitutes matter, whether organic or inorganic, is dependent on discreet entities that, gathered together form aggregates of new units which interact in support of one another and become an unfolding event, in constant motion,

with ever-increasing complexity and intricacy of their organisation [Ibid.].'

Thus the purpose of reconvergence is to reveal to us that evolution and economic transformation are interchangeable and that the entire ecosystem of the interweaving of man, nature, and the universe constitutes a single living totality. In this totality of things, man seeks his legitimate position in the unity and cosmic scheme of things. But in doing this, Ashen argues, we have to realise that we are in a situation of 'extreme darkness', in which there is a moral atrophy and destructive radiation within us as we watch the collapse of values hitherto cherished-'but now betrayed'. According to her: 'Science now seems to be telling us to question its previous premises and tells us not what *is* but *what ought to be*; prescribing in addition to describing the realities of life, reconciling order and hierarchy.' In other words, science has lost its way with its reductionist bent and needs to be rediscovered to the needs of restoring what has been destroyed.

In terms of agricology that we advocate, this reconvergence calls on us to retrace our understanding of the practise of agriculture from the wisdom of ancient times, from which a holistic science emerged along with the growth of agriculture and animal husbandry. We have already referred to the origins of agriculture in Egypt, in which Osiris and Isis discovered corn and other grains and spread the knowledge of agriculture throughout the world. The modern concept, *Ceres*, is directly linked to these ancient conceptions of agriculture. As we told the story from James Frazer's *Golden Bough*, the myth goes that Osiris reclaimed Egypt from savagery and gave the Egyptians laws which taught them to worship the gods. In this process Isis, the sister-wife of Osiris, discovered wheat and barley growing wild.

Osiris then introduced the cultivation of these grains among his people who abandoned cannibalism and took to the corn diet. Osiris is also said to have been the first to gather fruit from trees and to train vines to poles and to tread the grapes. Eager to communicate these discoveries to all mankind, Osiris is reputed to have left the whole government of Egypt to his wife, Isis, so that he could travel all over the world to diffuse the blessings of civilisation (including agriculture) to the rest of mankind wherever he went. The myth goes that wherever he went and found the soil unsuitable and the climate harsh to the cultivation of vine, he would teach the local inhabitants to console themselves for the want of wine by brewing beer from barley instead [Ibid: 436-37]. This myth is coherent and believable because it later translates into real history of Egypt and Europe and Asia of the time.

The myth also goes that Osiris was known as the Corn-god so that one of his personifications was corn, which was said to die and come to life again every year. This fact demonstrates that even at this early stage, the *circular nature of agricultural production* was part of the mythology of human reproduction that became the basis for human action in agriculture. The myth was, in fact, manifested into reality through rituals that celebrated in festivals the birth and resurrection of Osiris as personified in the corn from which the Christian belief of resurrection was drawn. The myth, therefore, later translated itself into a spiritual and religious system, which however later got 'secularised' by the advocates of capitalism.

It is not modern science or university colleges of agriculture that have produced the 'scientific knowledge' for perpetuating such ancient knowledge and wisdom. It has been through the 'living word' carried by the memory of every generation linked to the past that humanity has been able to pass on the inherited heritage from ancient times. It was through memory that linked our actions to a sustained living based on that knowledge that continues to perpetuate IKS as a living knowledge system. This is why the epistemology of Afrikology holds that for humanity to restore a holistic life that is not fragmented by the modern 'scientific epistemology' that separates the body and the mind and the individual from the community, we must move towards a holistic, combined and transdisciplinary approach to knowledge production that is moral and ethical, and that is at the same time linked to ancient knowledge and wisdom. We have in fact noted that, in their activities, small farmers engaged in organic and 'sustainable' agriculture have used such knowledge and continue to believe in a spiritual world that links them and their actions to the cosmic forces. Therefore, for them, a call for a return to the spiritual and moral world that can restore our relationship with nature, which they believe is sacred, is a call in the right direction.

This is why Afrikology is the epistemology that can help to lead us in this direction. This is because Afrikology is a re-assertion of the divine origin of the *word* (which the Greeks called *logos*) from which we derived our languages and the written scripts from which *all meanings* emanate. This is the epistemology of the origin of the universe as perceived by the humans and as documented in the *Memphite Theology* of Ancient Egypt, which the British stole from the Egyptian archives and now lies torn up in the British Museum. According to that epistemology the *naming* of things, which was done through the *uttering of the word* by the tongue was brought about by the senses of the body – the eyes, the nose, the ears, the tongue being linked to the heart to which they reported.

Thus, according to this original epistemology, it is the heart that named what the senses had experienced (by seeing, feeling, smelling, and/or hearing) which was then passed on to the tongue that uttered what the heart had named, which became the 'Word'. Hence, it is the concepts that the heart perpetually creates from these sensual human experiences that *empirical knowledge* is generated. Thus the 'word', as originally conceived by the heart and uttered by the tongue through human languages, is what constitutes human knowledge. Indigenous Knowledge Systems are therefore a product of this epistemology everywhere in the world. This knowledge is perpetually reproduced through human communication, dialogue and conversations as part of the daily human activities and interactions to produce a self-created world. Thus, through all human languages, knowledge is produced through the heart and then passed on from person to person through daily communication and speech by the tongue. That is why human speech and dialogue are of a divine origin because from their essence they emanate from the heart, which carries all concepts and meanings, which are divine as products of the heart.

But this original epistemology was contradicted by the modern 'scientific' or Cartesian epistemology, which separated the mind from the heart and the body from the mind in order to create its own language called the *logico-mathematical language*, through which 'scientific knowledge' is generated and communicated. It is this artificial language through which the academic 'disciplines' including the 'natural sciences' were created and developed. Hence, it was through this artificial logico-mathematical language that 'science' was able to arrive at the conclusion that it is only by 'science' 'improving' the seed that humanity can produce enough food to feed the expanding population. But in fact the motive behind the 'scientific agriculture' was not to feed the world but to make superprofits for agribusiness in the name of 'progress'.

This kind of knowledge has proved itself to be false, as we have seen, and this has been proven by small organic farmers through their indigenous knowledge. This proves that indigenous knowledge is superior to modern 'scientific knowledge', which in turn proves that the original epistemology of the heart, being at the centre of all knowledge, is the basis on which we can rejuvenate the ecosystems through a reconvergence of nature and the human beings. Since this knowledge is perpetually generated through their languages linked to the original 'living word', it stands to reason that the original knowledge is still valid and can be validated within itself. This is what we call the epistemology of Afrikology, which must inform IKS and agroecology if the agro-ecosystem is to survive.

Therefore, whatever is seen by the eyes, heard by the ears and smelled by the nose is perpetually reported to the heart, which ponders over these experiences, conceives their meanings and finally gives them names – which is the meaning attached to the 'word'. This process goes on every moment the human beings experience reality through their senses, memory and their actions and interactions. From this it follows that it is only the heart which is the seat of the truth and this truth can only be validated by dialogue and conversations between the hearts of the people who create and speak languages. The experience we have outlined above of the organic farmers engaged in different forms of agroecology proves this epistemology to be correct, hence the concept *agricology*, standing for both the knowledge (*Afrikology*) and the practises *(agroecology)*, can be the holistic basis for the rejuvenation of nature and production on a new spiritual and moral basis.

In Conclusion

The struggle to reclaim the land, repair the ecosystem and bring conventional capitalist 'scientific' agriculture to an end is a momentous revolutionary action which must be undertaken mainly by small farmers and the holistic scientists who are currently engaged in agroecology, conventional farmer's organic farming systems and other forms of alternatives that constitute the practise of 'sustainable agriculture'. These alternatives include: permaculture; the no-tillage systems of grain cultivation practised in Japan; bio-intensive agriculture utilising the 'double digging' method; bio-dynamic farming; sustainable systems using perennial varieties as well as systems employing companion planting, etc.

It is a struggle for survival of the universe and the human race. The war must be fought on the real ground not by mimicking nature in order to revive and reclaim the soils and the seeds; but it must be a battle that first and foremost restores nature by *replication*. In the case of the Indian organic farmers, one of their central principles, unlike those of the agroecology scientists, was not to mimic but to replicate nature:

'The principal idea was that the more closely you were able to replicate the practises of the forest, the richer the conditions of the soil and soil life and the better the crop you anticipate. The greatest consequence was the recognition that by doing this, one could also cut oneself from dependence on companies, banks, seed suppliers, extension agencies, equipment manufacturers and university scientists [Dharmitra, T. K., 2009].'

Thus, it is through both *regenerating* nature and replicating its products such as forests, that one could also restore the farmers' independence and freedom to live and practise what was at the same time supportive of nature. The farmers had not just talked; they had also demonstrated that they could walk the talk. They adopted several techniques of replicating in their fields by creating rich littered floors of natural forest and this became a paradise for soil fauna. They created mulches which could protect their soils from the sun and at the same time ensuring the continued existence of the beneficial microbes underneath. They had created an extensive expertise in *vermiculture* that had developed across the country and the wide range of recipes through which farmers were increasing the population of beneficial microbes and earthworms in their soils. The farmers' activities were proved to be a revolutionary achievement that had not required AK47 guns or police or agricultural scientists and extension workers or even subsidies from governments. They demonstrated that when nature was looked after well, it would also be overwhelmingly generous in rewarding those who cared for it. They demonstrated that farming on the basis of natural principles from well-tested indigenous knowledge systems was the only trusted way of keeping all of us alive.

The analogue of the forest was also able to remediate the problem of water contamination caused by the extensive use of chemical fertilisers and insecticides. They found out by experience that the concentration of nitrates, nitrites, chloride and potassium that was used to back up chemicalised agriculture was responsible for the damage of the soils and the water below it. This had led to different kinds of diseases including abortions in women, as well as cattle, goats, pigs and sheep, which were linked to contaminated water supplies. The remedy was to use bioremediation methods to reduce the concentration of nitrate and nitrite in the soil. This included the use of both microbes and plants. The principal means was the restoration and replication of the vegetation in the micro watershed around a drinking water well. Over a period of six to seven years, deep-rooted plants were established through these bioremedial means, forming a root web below the surface to draw up contaminants. The pilot area was then converted into a production area for both trees and annual crops including vegetables. Within four years, there was a decrease of nitrate and nitrite in the soil and water. The biodiversity created brought birds, butterflies and reptiles to the area.

All these experiences go to disprove mainstream ideologies framed as theories, which have disempowered communities through state policies and private interest interventions supported by the States. For instance, Elinor Ostrom [2009], winner of the Peace Prize in Economics in 2009, has given evidence from marginal communities in Latin America, Asia and Africa to disprove Hardin's theory of 'tragedy of the commons'.

Garret Hardin, in his 'Tragedy of the Commons' thesis published in the magazine, Science, in 1968, had argued that the practise of opening up the commons of pasture to all was bound to result in a 'tragedy' and environmental degradation unless the State intervened to regulate its use, or unless private enterprise assumed control, for only then would its exploitation be carried out in a 'rational' manner. In fact, Hardin had approached the question from the standpoint of an idealised model of the 'rational herder' and not the 'real herder' on the ground where, according to his model, each herder receives a direct benefit from his own animals and suffers delayed costs from the deterioration of the commons when others' cattle overgraze. According to the model, each herder is motivated to add more and more animals to the commons because he receives the direct benefit of his animals and bears only a share of the costs resulting from overgrazing. Hardin concluded:

'Therein is the tragedy. Each man is locked into a system that compels him to increase his herd without limit – in a world that is limited. Ruin is the destination toward which all men rush, each pursuing his own best in a society that believes in the freedom of the commons [Ibid: 2].'

In order to disprove Hardin's model, Professor Ostrom referred to a number of models drawn from a similar theoretical angle, including that drawn from Hobbes' 'state of nature' prototype of the tragedy of the commons, to draw our attention to the fact that if the only 'commons' were a few grazing areas or fisheries grounds, the tragedy of the commons would have been of little general interest. But this was not the case. In fact, she pointed out that Hardin had used his thesis on the tragedy of the commons to advance metaphorically a completely different argument – that of overpopulation. Following those examples arrived at metaphorically, Hardin had used all kinds of schemes to imagine all kinds of 'tragedies' such as the Sahelian famine of the 1970s; the firewood crisis throughout the Third World; the organisation of the Mormon church; urban crime; the problem of acid rain; and even the public-private sector relationships in modern economies to make his argument [Ibid: 3]. From these metaphorical models, other theories were put forward to justify external interventions such as the 'prisoner's dilemma game' and 'the logic of collective action'.

But Ostrom argued that as long as individuals using 'common pool resources' were viewed as 'prisoners', it was only state policy prescriptions that could 'rationally' address those imagined 'dilemmas'. Hardin's model was intended to impose a constraint on individuals to find *alternative* solutions to their problems. Ostrom pointed to the dangers of using models arrived at metaphorically. She pointed out:

'When models are used as metaphors, an author usually points to the similarity between one or two variables in a natural setting, and one or two variables in a model. If calling attention to similarities is all that is intended by the metaphor, it serves the usual purpose of rapidly conveying information in graphic form [Ibid: 7-8].'

However, the three models she presented had been used metaphorically for other purposes: She added:

'By referring to natural settings as 'tragedy of the commons', 'collective-action problems', 'prisoner's dilemma', 'open-access resources', or even 'common property resources', the observer frequently wishes to invoke an image of helpless individuals caught in an inexorable process of destroying their own resources [Ibid: 7-8].'

This conclusion leads to the conclusions and policy prescription that creates the impression that only the Leviathan is the way, and that the State or 'iron governments' or even military governments are the only ones that have the ability to use coercive power to save those 'imprisoned' in 'irrational collective activity'. The other 'solution' is the call for the privatisation of the commons or socialising them as the alternative. In the absence of the State or the private sector, it is prescribed that public agencies or even international authorities be brought in to deal with the 'tragedies'. This advice has been followed by 'progressive' governments in the Third World nationalising or socialising common resources from small farmers such as forests, fishing grounds, grazing lands, and other natural resources such as oil. Yet the results of such nationalisations of communal forests have been demonstrated to be disastrous in the case of India, Thailand, Nepal and Niger [Ibid: 23]. Many of the idealised prescriptions are metaphors advocating oversimplified 'institution-free' institutions. Ostrom continues:

> 'An assertion that central regulation is necessary tells us nothing about the way the central agency should be constituted, what authority it should have, how the limits on its authority should be maintained, how it will obtain information, or how its agents should be selected, motivated to do their work, and have their performances monitored and rewarded or sanctioned. An assertion that the imposition of private property rights is necessary tells us nothing about how that bundle of rights is to be defined, how the various attributes of the goods involved will be measured, who will pay for the costs of excluding non-owners from access, how conflicts over rights will be adjudicated, or how the residual interests of the right-holders in the resource system itself will be organised [Ibid: 22].'

Hence relying on metaphors as the foundation for policy measures can lead to results that are substantially different from those presumed by the models. Moreover, the predetermined premise and assumption that individual 'rational choices' lead to 'irrational outcomes' challenges the generally held fundamental faith that rational human beings can achieve rational results. Such an assertion results in reposing rational decisions in the State or private enterprises and not in individuals. This is a paradox that, in the case of the 'prisoners' dilemma', suggests that it is impossible for rational creatures to cooperate, which is fundamentally flawed. This directly bears on issues of ethics and political philosophy, which threatens the foundations of the social sciences [Campbell, 1985: 3]. Professor Ostrom warns that 'the intellectual trap' of relying entirely on metaphorical models to provide the basis for policy analysis is that scholars then presume that they are 'omniscient observers' who are able to comprehend the essentials of complex, dynamic systems and how they work, and thereby create stylised descriptions of some aspects of those systems. With the false confidence of presumed omniscience, scholars feel comfortable in addressing proposals to governments that are conceived in their models as omni-competent powers able to rectify the imperfections that exist in the field settings [Ibid: 215].'

Ostrom points out that the problem with this kind of approach is that most current analyses of 'common pool resources' is that they focus on a single or one-dimensional operational level of analysis. At this level, one assumes that both the rules of the game and the physical, technological constraints are statically given and that they will not change during the time frame of the analysis. But the fact of the matter is that actions of individuals in an operational situation are dynamic and they directly affect the physical world. These actions include such decisions as withdrawing resources from the common pool; the transforming of inputs into outputs; the exchange of goods; as well as appropriation and provision problems occurring at an operational level.

There is therefore no fixed and unchangeable state in regard to technology and institutional rules. Both technology and rules are changing over time as decisions are made by individuals. In practise, the analysis of technological changes has proved to be more difficult than changes of production and consumption decisions within a fixed technological framework [Nelson and Winter, 1983]. Analysis of institutional changes has also proved more difficult than the analysis of operational decisions within a fixed set of rules [Ostrom, Ibid.].

Yet, the examples of the destructive role science and technology have played in agriculture demonstrate that the static models of the transfer of technology for purposes of 'catching up' of the poor societies with the more 'developed ones', is erroneous. Instead, the evidence proves that it is the so-called irrational indigenous knowledge systems that have emerged to prove the mainstream 'scientific' systems to be wrong in most respects in agriculture. The knowledge was maintained by the farmers' self-reliant decisions without committing aggression against anyone. The knowledge has also brought about a restorative and self-reliant culture all round. The knowledge has strengthened the interconnected self-reliance of the soils, water, and seeds, which reproduce themselves when the rules of the game in the indigenous knowledge systems are observed.

The organic farming experiences in Asia, Africa and other parts of the world have shown that small farmers throughout history have been, and continue to be, the best replicators of seed and plant life. This is because anyone who can successfully grow any crop is equipped to save its seed and replicate or multiply it through reproduction. This is why the ancient civilisations in Egypt likened this process to the birth, death and resurrection. For millennia, ever since the plants were domesticated in the ancient world, seeds have been inseparable from the farming of the communities living on the 'commons'. According to Kundaji:

> 'All agricultural biodiversity that is our heritage has been maintained and developed in such farms and these unknown farmers are our real seed breeders – claiming no property rights; no royalties; no acknowledgements. This situation started changing when the first 'experts' were brought in to 'improve' crops and make agriculture 'profitable' (for companies). Once these seeds were released and pushed on farmers, packaged with accompanying inputs, it was the beginning of the end of traditional agriculture and its plant wealth [Kundaji, 2009: 15].'

Thus, there can be no doubt that the models based on reductionist sciences, metaphoric models and fixed input-output technologies should be seen as destructive of nature. We also can come to the conclusion that it is only the small farmer, small fisherman and herdsman that can save the 'commons' if they are guaranteed the rights over their lands and properties. The dispossession of the small farmer, which has led to the emergence of large agribusiness, has led to the destruction of the 'commons' and turned them into dustbowls of large-scale farming, which have turned into disasters. Yes, large profits have been made by individuals, but at what cost?

REFERENCES

Altieri, M. A. [1994]: *Biodiversity and Pest Management in Agrosystems*, Hayworth Press, New York.

Altieri, M. A. [1999]: 'Agroecology: Principles and Strategies for Designing Farming Systems', University of California, Berkeley.

Alvares, C. [2009]: 'The Problem of Degraded Soils, Insects and Weeds', in *Third World Resurgence*, No. 230, October 2009, Penang.

Alvares, C. [2009]: 'Organic farmers can feed the world', in *Third World Resurgence*, No. 230, October 2009, Penang.

Anshen, R. N. in Chomsky, N. [1986]: *Knowledge of Language: Its Nature, Origin, and Use*, Praeger, New York.

Assmann, J. [1996]: *The Mind of Egypt: History and Meaning in the Time of the Pharaohs*, Harvard University Press, USA.

Badgley, C.; J. Moghtader; E. Quintero; E. Zakem; M. Chappell; K. Aviles-Vazquez; A. Samulon; and I. Perfecto [2007]: 'Organic Agriculture and the Global Food Supply' in *Renewable Agriculture and Food Systems*, 22 (2).

Borlaug, Norman E., 'The Impact of Agricultural Research on Mexican Wheat Production', in *Transactions of the New York Academy of Science*, 20 (1958) 278-295.

Borlaug, Norman E., 'Wheat Breeding and Its Impact on World Food Supply': Public lecture at the Third International Wheat Genetics Symposium, August 5-9, 1968. Canberra, Australia, Australian Academy of Science, 1968.

Both Ends [2007]: 'Adapting to Climate Change: How local experience can shape debate': *Both Ends Briefing Paper*, August 2007, Amsterdam.

Berlin, I. [1979]: *Against the Current: Essays in the History of Ideas*, Princeton University Press, Princeton / Oxford.

Bernal, M. [1987]: *Black Athena: The Afro-Asiatic Roots of Classical Civilisation*, Volume One, Free Association Books, London.

Chossudovsky, M. [2010]: 'Global Poverty and the Economic Crisis' in Chossudovsky, M. and Marshall, A. G. [2010]: *The Global Economic Crisis: The Great Depression of XXI Century*, Global Research, Montreal, Quebec.

Diop, C. A. [1980]: *Civilization or Barbarism: An Authentic Anthropology*, Lawrence Hill Books, Chicago.

Dharmitra, T. K. [2009]: 'From Industrial Agriculture to Agro-Ecological Farming – A South African Perspective', *Working Paper Series No. 10*, ECSECC.

ECT Group [2010]: *The New Biomassters: Synthetic Biology and the Next Assault on Biodiversity and Livelihoods* [2010].

Engdahl, F. W. [2007]: *Seeds of Destruction: The Hidden Agenda of Genetic Manipulation*, Global Research Press, Montreal, Quebec.

Finch, C. V. and C.W. Sharp [1976]: *Cover Crops in California Orchards and Vineyards*, USDA Soil Conservation Service, Washington.

Frazer, J. [1996]: *The Golden Bough*, Penguin Books, Harmondsworth, Middlesex.

Fuller, S. [2010]: *Science: The Art of Living*, Acumen, Durham.

Fuller, S. [1997]: *Science*, University of Minnesota Press, Minneapolis.

Friends of the Earth [2009]: *Africa-Up for Grabs: The impact and scale of land grabbing for biofuels*, London.

Jackson, T. [1996]: *Material Concerns: Pollution, Profit and Quality of Life*, Rutledge, London.

Jhamtani, H. [2010]: 'The Green Revolution in Asia: Lessons for Africa', in *Third World Resurgence*, Penang, Malaysia, Issue No 223: 'Rethinking Agriculture: New Challenges, a New Development Agenda'.

Juma, C. [1989]: *Biological Biodiversity and Innovation*, African Centre for Technology Studies, Nairobi, Kenya.

Galton, F. [1904]: 'Eugenics', *The American Journal of Sociology*, Volume X; July, 1904; Number 1.

Gleissman, S. R. [1998]: *Agroecology: Ecological Processes in Sustainable Agriculture*, Ann Arbor Press, Michigan.

GRAIN [2009]: *Land Grab or Development Opportunity: Agricultural Investment and International Land Deals in Africa*, London.

Harvey, D. [2003]: *The New Imperialism*, Oxford.

Harvey, D. [2005]: *A Brief History of Neoliberalism*, Oxford.

Hoffmaister, J. [2009]: 'Resilience: More than a trendy word', in *Third World Resurgence*, No. 223, October 2009, Penang.

IAASTD [2008]: *Agriculture at a Crossroads: International Assessment of Agricultural Knowledge, Science and Technology for Development*, Island Press, Washington DC: www.agassessent.org.

Kundaji, D. [2009]: 'Farmers as seed breeders', in *Third World Resurgence*, No. 230, October 2009, Penang.

Loudon, J. C. [1825]: *Encyclopaedia of Agriculture*, Longman, Hurst, Rees, Orme, Brown, and Green, London.

Lovins, A. [1977]: *Soft Energy Paths*, NY, Harper Colophon.

Mae-Wan Ho, Harmut Meyer and Joe Cummings [1998]: 'The Biotechnology Bubble' in *The Ecologist*, Vol. 28, No. 3, May-June, 1998, London.

Mae-Wan Ho [2009]: 'Organic Cuba Without Fossil Fuels' in *Third World Resurgence*, Issue No. 216, March 2009, Penang.

Mafeje, A. [1988]: 'The Agrarian Question and Food Production in Southern Africa', in Prah, [1988]: *Food Security Issues in Southern Africa*, Institute of Southern African Studies, The National University of Lesotho, Southern African Studies Series No. 4.

Marx, K. [1976]: *Capital, Vol. I*, Universal Publishers, New York.

Marx, K. [1956]: *Capital, Vol. II*, Peoples' Publishing House, Moscow.

Marx, K. [1973]: *Grundrisse*, Penguin Books, Auckland.

Mayet, M. [2010]: 'Africa's Green Revolution rolls out the Gene Revolution', *Third World Resurgence*, Penang, Malaysia, Issue No. 223: 'Rethinking Agriculture: New Challenges, a New Development Agenda'.

Murphy, C. [2000]: 'Cultivating Havana: Urban Agriculture and Food Security in the Years of Crisis', May 2000.

Nabudere, D. W. [2002]: 'The Interface between Traditional Systems of Governance and Contemporary State Systems in Africa', paper written for UNDP, Mbale, Uganda (unpublished).

Nabudere, D. W. [2009a]: *The Crash of International Finance Capital and its Impact on the Third World*, New edition published by Pambazuka Press, London, and Fountain Publishers, Kampala.

Nabudere, D. W. [2009b]: *The Global Capitalist Crisis and the Way Forward for Africa*, SEATIN, Kampala.

Nabudere, D. W. [2011]: *Afrikology, Philosophy, and Wholeness: An Epistemology*, Africa Institute of South Africa, Pretoria.

Odhiambo, A. [2007]: 'AGRA takes Certified Seeds to Farmers in War on Hunger', *Business Day*, Nairobi.

Okot, p'Bitek [1971]: *African Religions in Western Scholarship*, East African Literature Bureau, Nairobi, Kenya.

Pretty, J. N. [1994]: *Regenerating Agriculture*, Earthscan Publications Ltd, London.

Pingali, P. L. and M. W. Rosegrant [1994]: 'Confronting the Environmental Consequences of the Green Revolution in Asia', *EPTD Discussion Paper No. 2*. http://www.ifpri.org/divs/eptd/dp/papers/eptdp02.pdf

Ostrom, E. [2008]: *Governing the Commons*, Cambridge University Press, Cambridge.

Ramose, M. B. [2002]: *African Philosophy Through Ubuntu*, Mond Books, Harare.

Ray, D. I. and P. S. Reid (editor) [2003]: *Grassroots Chiefs in Africa and the Afro-Caribbean Governance*, University of Calgary Press.

Robinson, W. I and J. Harris, [2000]: quoted in Chossudovsky, M. and A. Marshall, [2010]: *The Global Economic Crisis: The Great Depression of XXI Century*, Global Research, Montreal, Quebec.

Sabbadini, S. A. [2010]: 'Quantum Physics, Cosmology and Spirituality', presentation made at Retreat 3 of the SARCHI Chair in Development Education at the University of South Africa (UNISA), Pretoria, on 23 November 2010.

Sumner, D. R. [1982]: 'Crop Rotation and Plant Productivity', in *Handbook of Agricultural Productivity, Vol. I*, CRC Press, Florida.

Tayob, T. K. [2010]: New Book on possible 100 per cent renewable energy by 2050 in *Third World Resurgence*, Issue 231-232 (Double Issue), Penang, Malaysia.

Reinjntjes, C. B. etc. [1992]: *Farming for the Future*, MacMillan Press Ltd, London.

Taylor, C. [1985]: *Human Agency and Language: Philosophical Paper 1*, Cambridge University Press, Cambridge.

Williams, C. [1989]: *The Destruction of Black Civilization: Great Issues of Race from 4500 BC to 2000 AD*, Third World Press, Chicago.

White, L. [1967]: 'The Historical Roots of Our Ecological Crisis', *Science*, Volume 155, pp. 1202-1207.

World Bank [2010]: *Rising Global Interests in Farmland: Can it yield sustainable and equitable benefits?* Report, Washington.

Vandermeer, J. [1995]: 'The Ecological Basis of Alternative Agriculture', *Annual Review of Ecological Systems* 26: 201-224.

Von Werlhof, C. [2010]: 'Globalisation and Neoliberalism: Is there an Alternative to Plundering the Earth?' in Chossudovsky, M. and A. G. Marshall [2010]: *The Global Economic Crisis: The Great Depression of XXI Century*, Global Research, Montreal.

Von Sertima [1999]: 'The Lost Sciences of Africa: An Overview' in Makgoba, M. W. [1999]: *African Renaissance*, Mafube, Tafelberg, South Africa.

Von Sertima [1989]: *Before Columbus*, New York.

Zarlenga, S. [2002]: *The Lost Science of Money; The Mythology of Money – the Story of Power*, American Monetary Institute, New York.

INDEX

Abrahamic narrative 44–45
Aduso philosophy 12, 192, 200–201
Africa,
  Green Revolution 24, 79–81, 84-89
  lessons from Asia 87-89 *see also*
  country concerned
*Africa – Up for Grabs: The Scale and
  Impact of Land Grabbing for Agrofuels*
  91
Africa Rice Centre (WARDA) 87
African Agricultural Technology
  Foundation (AATF) 85
African Development Bank (ADB) 86
African Rice Initiative (ARI) 86–87
Afrikology concept 198, 204, 206
agribusiness 24, 60–61, 65, 69–73, 75–76,
  79–80, 86–87, 89–90, 99, 107–109, 112,
  -116–117, 121, 130, 133, 143, 149, 191
agriceutical super monopolies 99, 112, 115
agricology 13, 89, 161, 184, 192–193,
  197–198, 203, 206
agricultural origins 18–23
Agriculture Development Council 106
Agriculture at a Crossroads 39, 47, 88–89,
  100, 103, 105, 134, 136, 140, 142, 167,
  169–170
agro-chemicals 88, 148, 154–155, 170–171;
  see also: chemical usage
Agro-Dealer Development Programme 82
agro-dealers 82–84
agroecology 170–178, 182–187, 206
agrofuel 91–95, 115
Agroils 97
Alliance for a Green Revolution in Africa
  (AGRA) 80–82, 84–86
Altieri, M.A. 175, 182–183
Alvares, C. 24, 190
American Population Control League
  62–63
Ancient Egyptian agricultural mythology
  18–20, 22, 203-204
Annan, Kofi 80
Anshen, Ruth Nanda 202–203
Argentina, soybean monoculture 71–73
Aristotle 47
Asia, Green Revolution 79–81, 87-88, 94
Asian Development Bank 58

Assmann, Jan 199
Bacon, Francis 48–49
Badgley, C. 186
Bayer CropScience 83
Berry, Thomas 14–15, 24
Bill & Melinda Gates Foundation 80–81,
  85
Biofuel Africa 92, 97
biofuels 16, 90–93, 96, 115, 126
biomass 16, 113–114, 121, 123-129, 137,
  172, 188
Biomass Power Association 125
Biosafety Project 85
biotechnology 149
Black virgin cult 20
Blood of a Nation 64
Borlaug, Norman Ernest 55, 57
Brazil, soybean monoculture 73
British Department for International
  Development (DFID) 88
Buffett Foundation 85
Bush, George H. W. 115
Bush, George W. 90
Capital (Das Kapital) 28
capitalism 14, 16, 26–30, 32–33, 37–39, 95,
  115, 132–134, 167, 184–185
Cargill 107
Carnegie Foundation 63- 64
cattle wealth 196–197
chemical usage 15, 24–25, 27–28, 31,
  53–54, 59–61, 66, 72, 74–75, 79, 81–82,
  105, 152, 154, 168, 188–189, 191, 207
Chicago Board of Trade (CBOT) 35
Chossudovsky, Michael 74
Citizens Network for Foreign Affairs
  (CNFA) 81, 83
climate change 14, 17, 52, 100–105, 113,
  127, 163, 165, 180
Coffee, Sugar and Cocoa Exchange (CSCE)
  35
collective property concept 149–150
Comite Pro Vida de Mexico 77
commodity markets 35–38, 90
Comprehensive African Agriculture
  Development Programme (CAADP) 81
Conservation Foundation 127
Consultative Group on International

Agricultural Research (CGIAR) 87
Costello, John 83
Croplife Foundation 83
Cuba, Green Revolution 152
    agricultural reorganisation 153–156
    Organponicos 156–157
da Silva, Lula 73
Dabholkar, Professor 188
Daewoo Logistics 97
Danford Centre 85
Darwin, Charles 51, 62, 63
Delta & Pine Land 76–78
Descartes, Rene 41–42, 49, 67, 161, 198,
    202, 205
Dharmitra, T.K. 213
Diageo 84
Diamond vs. Chakrabarty (USA) 78
Diaz, Miguel 78
Diop, Cheikh Anta 16, 20, 134
Dow AgroScience 83
Dow Chemicals 54
Dravidians 23
DuPont 54, 84, 125
DuPont Crop Genetics Research 85
DuPont Crop Protection 81, 83
Earth Institute (Columbia University) 82
Ebukalin, Sam 201
ecosystem exploitation 14–17, 23–25, 31,
    37–38, 52, 59, 72, 74, 80, 88, 92–93, 95,
    100, 103, 105, 127, 130
ECT Group 116, 124, 127
Encyclopaedia of Agriculture 21
energy systems 124-125, 193–196
Engdahl, F.W. 65-66, 71, 75, 106-107,
    127–128
Environmental Management Group
    (EMG) 130
Essay on the Principles of Population 62
ethanol 90–91
Ethiopia, GMO seeds 73, 79
eugenics 62–66, 77, 127
European Food Safety Agency (EFSA) 118
farmer innovation 139–140, 144–146, 150,
    165–166, 190–191
financial services' role 34–37, 85-86, 90, 95
Food and Agriculture Organisation (FAO)
    35, 96, 98, 106, 154

Food and Drug Administration (FDA) 118
food issues 56, 87, 89, 96–99, 106,
    109–110, 112, 147–148, 152–156, 163,
    164–165, 181, 187
Ford Foundation 57, 61, 75–76, 108
Framework for African Agricultural
    Productivity (FAAP) 81
Frazer, Sir James 18-19, 21, 203
Friends of the Earth 91, 96
Fuller, Steve 40–41, 43–45, 66
Galten Global Alternative Energy 97
Galton, Francis 62–63
garden city concept 156–160
Genetic Use Restriction Technology
    (GURT) 75–76
Genetically Modified Organisms (GMOs)
    13, 17, 24, 61, 71–76, 79, 81–82, 84–86,
    88, 100, 117–118, 126
genetics / genetic engineering 51, 62–71,
    75, 77–78, 82, 122, 127, 149, 170, 172
Gethi, James 86
Glass-Steagall Act (USA) 36
Global Coalition of Social Forces for
    Another World 130
Global Network of Physicians and
    Scientists for Responsible Application of
    Science and Technology 118
Gold Star Farms 97
Goldberg, Ray 65
Goldman Sachs Bank 109
GRAIN 87, 95–96
Green Revolution origins 53–56; See:
    country concerned
Haber Bosch process 126
Hansen, Jim 101
Hart, Tim 146
Hardin, Garret 181, 208–209
Harris, Jerry 132
Harvard Economic Research Project 108,
    112
Harvey, David 95
Hegel, Georg Wilhelm Friedrich 45
Hercules Powder 54
Herskovits, M.J. 197
Hitler, Adolf 64
Ho, Mae-Wan 69, 156
Hobbes, Thomas 208

Hoffmaister, J. 162–163, 165
How Wall Street and Washington Betrayed America 36
Howard, Albert 24
Howard, *Sir* Ebenezer 158
India, GMO seeds 76;
　Green Revolution 24, 55–60, 106–107, 187
Indigenous Knowledge for Development Program 141
indigenous knowledge systems 9, 11–13, 140–144, 146–150, 161–162, 167, 171, 178–179, 184–187, 192, 198, 200–201, 204–205, 207, 211 *see also* knowledge systems
Integrated Pest Management (IPM) 168
Intellectual Property Rights 143, 148–150
Intergovernmental Panel on Climate Change 101, 103
International Assessment of Agricultural Knowledge, Science, Science and Technology for Development 39
International Fertiliser Development Centre (IFDC) 81
International Food Policy Research Institute 94
International Fund for Agricultural Development (IFAD) 85, 96, 98
International Institute for Environment and Development (IIED) 96
International Maize and Wheat Improvement Centre (CIMMYT) 56
International Monetary Fund (IMF) 79, 137
International Movement for Ecological Agriculture 130
International Planning Committee for Food Security 98
International Rice Research Institute (IRRI) 70, 88
International Union for the Protection of New Varieties (UPOV) 168
Isis 18–20, 22, 74, 142, 203
Jackson, Tim 53
Jatropha Africa 97
Johnson Foundation 130
Jones, Monty 86, 103

Jordan, David Starr 64
JP Morgan Chase Bank 71
Juma, Calestous 48–49
Kapiza, Dinnah 83
Kenya, agro-dealing, 83;
　genetically modified maize 86
Key Farmers' Trainers 192
Kilimo Trust 84
Kimminic Corporation 97
Ki-moon, Ban 106
Kissinger, Henry 54
knowledge systems 135–139 *see also* indigenous knowledge/systems
Kuhn, Thomas 40
labour issues 30, 32–33, 151
land degradation *see* ecosystem exploitation
Land Grab or Development Opportunity? Agricultural Investment and International Land Deals in Africa 96
land grabbing 16, 91–93, 95–99, 113, 133
land rehabilitation 187–191, 207–208
Land Research Action Network 98
landed property concept 28–29, 32, 151
Latin America, Green Revolution 80–81, 94
Lenin, Vladimir 33
Linnaeus, Carl 50
Loudon, John Claudius 21
Lovins, A. 193
Luswazi, Professor 13
Ma'at philosophy 199–200
Machu Picchu agricultural achievements 40–41
Mafeje, A. 10, 30
Malawi, agro-dealing 83
Malthus, Thomas 62–63, 156, 181
Marx, Karl 26, 28–29, 32–33, 111
Mayet, Mariam 81
Mbaabu, Anne 84
Memphite Theology 204
Mendel, Gregor 51, 67
Menem, Carlos 71
Mesopotamia 20
Mexico, Green Revolution 55–56, 80, 107
Milani, T. 159–160, 194
Mitunguu Millers Ltd 83–84
molecular biology 65–66, 70, 115

monoculture 51, 60, 71-73, 81
Monsanto 13, 61, 71–73, 75–76, 78, 81–86, 88, 109
Mouton, Johann 146
Nabudere, Dani Wadada, obituary 8-10
nanotechnology 123–124, 129, 170
National Agricultural Research Organisation (NARO) 192
National Agricultural Research Systems 136
National Microfinance Bank 84
National Security Memorandum – NSSM 200 54
National Semi-Arid Resources Institute (Nassari) 192
National Aeronautics and Space Administration (NASA) 101
Nelson, Charles 22
New Partnership for Africa's Development (NEPAD) 81
New Rice for Africa (NERICA) 86-87
Newton, Sir Isaac 49, 51
Niger, agro-ecological management 163–165
Ohio State University 13
Organisation for Economic Co-operation and Development (OECD) 117
Osiris 18–20, 22, 74, 142, 203–204
Ostrom, Elinor 208–211
Overseas Development Institute (ODI) 180
Pan-African Non-petroleum Products Association (PANPP) 92
p'Bitek, Okot 199
peasant farmer displacement 151–153 see also small-scale farming
Pioneer HiBred International 85, 109
Pius IX, Pope 48
Planned Parenthood Federation of America 62
Popenoe, Paul Bowman 64
Popular Knowledge of Women's Initiative (PKWI) 11, 192-193, 200-201
population control 62, 64, 77–78, 127–128
Precautionary Principle 129–130
Promoting Local Innovations (PROLINNOVA) 145

Ramose, Professor 161–162
Rapid Rural Appraisal 146
Ray, D.I. 181
Reagan Administration 61, 69, 108, 112
Regal, Philip 68
Report on India's Food Crisis and Steps to Meet It 56
Rice 70, 86-88
Right to Food Conference 98
Rising Global Interests in Farmland: Can it Yield Sustainable and Equitable Results 91
Robinson, William J. 132
Rockefeller Foundation 56, 62–63, 65–66, 68–71, 77, 80, 107–108, 112, 127
Rockefeller, Laurance 127
Rockefeller, Nelson 55
Rockeffer Standard Oil 54
Roman agricultural mythology 19–20
Roosevelt, Franklin D. 36
Royal Society of Canada 118
Rural Code (Niger 1993) 164
rural/urban divide 151–160
Sachs, Jeffrey 82
St Francis of Assisi 48
Sandrolini, Christopher 78
Sanger, Margaret 62, 64
Scanfuel 97
scientific and technological issues 39, 42–52, 87, 211
sedentarisation 39
small-scale farming 70, 82–83, 87–88, 100, 106, 109–111, 131, 139, 153, 161–162, 180, 186–188, 190–191, 205, 207, 212 see also farmer innovation; peasant farmer displacement
South African Council for Scientific and Industrial Research (CSIR) 85
Southern Agricultural Growth Corridor of Tanzania (SAGCOT) 84
Stanbic Bank 84
Standard Oil Trust 108
structural adjustment policies 79, 137
sustainability concept 169–170, 176–177, 186, 192
Syngenta Crop Protection 83-84
synthetic biology 113, 115–123, 129, 170

Tamil Nadu Organic Farmers' Market 187, 189

Tanzania, breadbasket project 84-85

Taylor, Charles 42

technology transfer model 46–47, 211

terminator technology 76–77

Thatcher Administration 61, 69, 108, 112

The Destruction of African Civilisation 179

The Golden Bough 18, 203

The Mind of Egypt 199

The New Biomassters: Synthetic Biology and the Next Assault on Biodiversity and Livelihoods 116

The Oeconomy of Nature 50

The Origin of Species 51, 62

They Came Before Columbus 23

Third World Resurgence Journal 185

To-Morrow: A Peaceful Path to Real Reform 158

Trade Related Intellectual Property (TRIPS) 69–71

traditional intellectual property rights 150

traditional knowledge *see* indigenous knowledge

traditional political systems 179–181

Transfer of Technology model 81, 89, 144

Ubuntu philosophy 161–162, 199

Uganda, genetically modified maize 86

UN Special Rapporteur on the Right to Food 187

Unilever 84

United Nations 98, 187

United Nations Environmental Programme (UNEP) 115

United States. Department of Agriculture (USDA) 76

United States National Renewable Energy Laboratory 114

University of Iowa 110

University of Makerere 13

University of Uppsala 124

University of Sussex 98

USAID 78, 84

Vatican, and biotech food 78–79

von Sertima, Ivan 23, 194

Wade, Kip 92

Walter Sisulu University 13

Wambug Florence 85

Warehouse Receipt System 84

Water Efficient Maize for Africa (WEMA) 85–86

Weaver, Warren 66

Wendorf, Fred 22

Williams, Chancellor 179–180

Winters, Clyde 23

World Bank 57, 79, 91, 94, 96, 136, 140–141, 146

World Health Organisation (WHO) 77

World Meteorological Organisation (WMO) 103

World Resource Institute 163

World Trade Organisation (WTO) 69, 71, 78, 113

Yara 84

Zarlenga, S. 196